Advances in Lectin Research   Volume 4

To Josephine
* 28. 12. 1990

Advances in Lectin Research

Volume 4

Edited by Hartmut Franz

Coeditors: Ken-ichi Kasai, Jan Kocourek,
Sjur Olsnes, Leland M. Shannon

With 40 Figures and 13 Tables

Springer-Verlag Berlin Heidelberg GmbH

Hartmut Franz
Staatliches Institut für Immunpräparate und Nährmedien
Klement-Gottwald-Allee 317–321, Berlin,
O - 1120, Germany

Ken-ichi Kasai
Faculty of Pharmaceutical Sciences,
Teikyo University
Sagamiko, Kanagawa 199–01,
Japan

Jan Kocourek
Dept. of Biochemistry Charles University Prague
Hrusicka 2515, 14100 Praha 4 – Spořilov 11,
Czechoslovakia

Sjur Olsnes
Norsk Hydro's Institute for Cancer Research
Montebello Oslo 3,
Norway

Leland M. Shannon
University of California
Riverside, California 92521,
USA

Advances in Lectin Research. Volume 4 / hrsg. v. H. Franz. –

ISBN 978-3-333-00584-3   ISBN 978-3-662-26751-6 (eBook)
DOI 10.1007/978-3-662-26751-6

1. Auflage
© Springer-Verlag Berlin Heidelberg 1991
Originally published by Springer-Verlag in 1991
Softcover reprint of the hardcover 1st edition 1991

Satz: Förster & Borries Satz-Repro-GmbH Zwickau

Authors

Prof. Dr. Henri Debray
Université des Sciences et Techniques de Lille Flanders-Artois
Laboratoire de Chimie Biologique
59655 Villeneuve d'Ascq Cedex, France

Dr. Christian Flemming
Institut für Biotechnologie
Permoserstr. 15, Leipzig, O - 7050, Germany

Prof. Dr. Dr. Hartmut Franz
Staatliches Institut für Immunpräparate und Nährmedien
Klement-Gottwald-Allee 317–321, Berlin, O - 1120, Germany

Dr. J. Montreuil
Université des Sciences et Techniques de Lille Flanders-Artois
Laboratoire de Chimie Biologique
59655 Villeneuve d'Ascq Cedex, France

Prof. Dr. Gerhard Uhlenbruck
Universitätskliniken Köln
Abteilung Immunologie
Kerpener Straße 15, Köln 41, W - 5000, Germany

From the Foreword to Volume 1

The years 1987/88 appear to be a suitable moment for the publication of "Advances in Lectin Research". The detection of ricin 100 years ago led to interesting results concerning sugar-binding proteins, now called lectins. Today two main lines exist in "lectinology". Firstly, lectins are used as multi-purpose tools (analysis and isolation of glycoconjugates, characterization and preparation of cells and microorganism, lectin histochemistry). Secondly, lectins are of interest concerning their physiological roles and other biological activities. Especially during the last 10 years, the latter aspects have been widely noticed. More and more the opinion is being accepted that lectins are glues of moderate affinity, making possible the contact of biologically cooperating glycoconjugates and cells bearing glycoconjugates.

The results of lectin research anticipated for the coming years should be of great importance not only for basic research, but also for diagnostic and even therapeutic application (adherence inhibition of bacteria and tumor cells, immunomodulation, lectin-mediated cytotoxicity, chemically modified toxic lectins) ...

"Advances in Lectin Research", as a collection of review articles, is not in competition with textbooks and congress proceedings ...

Foreword to Volume 4

Following the now proved principles of the former volumes of "Advances in Lectin Research" we present in volume 4 again articles covering the diversity of present lectinology. The editor hopes for further comments and for advices. He has to thank both the Verlag Gesundheit GmbH and the Springer-Verlag, the coeditors and again Mrs. Margitta Hintz, Dr. Rolf Wachowius, Heinz Zorn and Mrs. Marianne Lobstein.

Hartmut Franz, Berlin

Contents

3   Lectin Affinity Chromatography of Glycoconjugates   51

H. Debray and J. Montreuil

9

Editorial

Lectinology on the Threshold of Its Second Century

Over the last few years the exciting beginning of lectin research in 1888/89 has been appreciated by a number of authors. Beyond doubt, the detection of ricin was a fateful hour, influencing the early period of immunology and even causing the origin of some of its branches (e. g., serotherapy, development of immunotoxins). The investigation of lectins has led to remarkable results mainly over the last two decades. Recently the author has attempted to tackle the balance of 100 years of lectin research (Franz 1990). On the threshold of the second century, the question "quo vadis lectinology?" might be justified. The rapid progress in biosciences and, hence, also in lectinology, complicates the answer to this question. On the other hand, in order to make the series "Advances in Lectin Research" as topical as possible, it would be very helpful to predict the main trends in development at least for the next decade. Therefore, I take the liberty to formulate my own ideas and, at the same time, realizing the risk of being corrected by the results of the next years.

1. I like very much the hypothesis of Jeremy Carver (Toronto) which states that the carbohydrate moiety of glycoconjugates contains a multiplicity of information with respect to their sugar sequence and three-dimensional structure. This information can be decoded by lectins. Carver's conception might help to better understand some of the physiological functions of lectins. Other in vivo effects are surely connected to the ability of lectins to act as "sugar specific glue" sticking cells together, as well as microorganisms and glycoconjugates. From a therapeutical point of view, a more sophisticated understanding of this phenomenon might help to prevent both microbial infections and the metastasizing of tumors.

2. In contrast to antibodies, lectins are characterized by rather broad specificity and relatively weak binding to receptors. However, in the presence of glycoconjugates with different sugar moieties lectins are blocked with regard to receptors of lower affinity. Receptors of a high affinity on the other hand, can compete with glycoconjugates to bind the lectin. Compared with antibodies, lectins are more promiscuous. The extent of this blocking phenomenon against lectins depends on the concentration and diversity of the glycoconjugates. The interchangeability of the lectin-sugar bond guarantees a steady state and increases at least the "specificity" of lectins in vivo. The modulation of this phenomenon could open new possibilities.

11

3. The exploration of the biological functions of lectins is more difficult than expected earlier. Fortunately, over the last 20 years we have obtained new insights concerning mainly microbial and animal lectins. Surprisingly, the functions of plant lectins are mostly unclear, although a great number of investigations have dealt with this problem. I suppose that at least some of the following questions will be answered in the next 10 years: Are plant lectins storage proteins? Are they "atavistic" molecules? Are lectins involved in the water balance of distinct plants? Do lectins influence the germination procedure? Moreover, no one can exclude completely new results in this field.

4. The investigation of the relationship between lectinology and immunology promises further findings. Some of the mediators of the immune system are lectins and lectins may play a key role in the development of targeted drugs. Furthermore, structures involved in lectins could be phylogenetic precursors of antibodies. Finally, I would like to draw attention to the structural similarity of toxic lectins like ricin and mistletoe lectin I (ML I) and class II major histocompatibility complex molecules. Ultimately, the question arises whether lectins and immunoglobulins are phylogenetically related. The comparison of molecules belonging to the immunoglobulin superfamily and distinct lectins (among the toxic lectins) could lead to new conceptions. Similar correlations might exist between lectins and enzymes.

5. Of interest is also the possible involvement of lectins in the modulation of nuclear functions and the biotransformation of lectins. Lectins will be increasingly included in biotechnology, and lectins modified by molecular cloning will give new information.

I would be happy and grateful to be able to critically review my predictions in a later volume of *Advances in Lectin Research*.

Hartmut Franz, Berlin

1 The Tridacna Lectins

Gerhard Uhlenbruck

1.1 Introduction: Historical Background

Ten years before we discovered the *Tridacna* lectins, the background work which led to this event, was begun by Prof. Otto Prokop in Berlin. He found and described the first well-characterized invertebrate lectins in the garden and in the land snail albumin gland, and also in the eggs of these snails (for review see Prokop and Uhlenbruck 1970). As a byproduct of those investigations, the snails galactans (May and Weinbrenner 1938) were rediscovered and a new group of protease inhibitors were detected, because the *Helix pomatia* lectin seemed to be indigestible (Uhlenbruck et al. 1971).

It is interesting to note that Prokop and his group in this connection also have described for the first time vertebrate lectins, after also searching for agglutinins in fish eggs (Prokop and Uhlenbruck 1970). However, at that time no one was aware of the fact that all these new precipitating, agglutinating and by carbohydrates inhibitable "antibody-like substances" were indeed "lectins". The opinion that lectins only occur in plants was in those days still a fundamental dogma, whereas we now know that these "protectins" all belong to one great family of lectins, widely distributed in bacteria, plants, invertebrates and vertebrates. In analogy to the snail eggs, also in fish eggs a new group of proteinase- inhibitors were described (Uhlenbruck et al. 1972).

1.2 The Discovery of Tridacna Lectins

As already mentioned, the vertebrate lectins were discovered in the course of investigations on fish eggs by Prokop and colleages. The galactan polysaccharides however, led much earlier to the discovery of vertebrate lectins, but in an indirect way, which clearly and evidently demonstrated the presence of a galactose-binding principle in the vertebrate liver. May and Weinbrenner (1938) injected large amounts of snail galactan into rabbits and found that whereas the amount of galactan disappeared from the blood, it accumulated selectively in the liver. In fact they made the same experiments, which Ashwell and his group made nearly 30 years later, but they could not interpret at that time the phenomenon, which was due to the hepatic galactose-binding lectins, and did not follow it up, as they were mainly interested in the metabolism of galactan (called in those days "galactogen").

As the *Helix pomatia* snail galactan was collected our lab during the preparation of the snail's lectin, we also examined the serological properties of this polysaccharide and found blood group H activity, due to the presence of L-galactose, detected with diffe-

13

rent catfish antisera (Baldo and Uhlenbruck 1973). The inclusion of the hemolymph of *Tridacna maxima* in our experiments was pure chance.
We had obtained and ordered it from a zoological catalogue in order to use up our budget for 1974. So we tested the crude hemolymph against our galactan preparation and found a very strong precipitation line, which was not due to any "anti-A" lectin contamination in the snail galactan sample (Uhlenbruck et al. 1975b). On the contrary, the lectin was located in the electrophoretic analysis of the *Tridacna* hemolymph, moving to the anode, and was not identical to other "anti-galactan" lectins (Uhlenbruck et al. 1975b). This is demonstrated in Fig. 1.1.

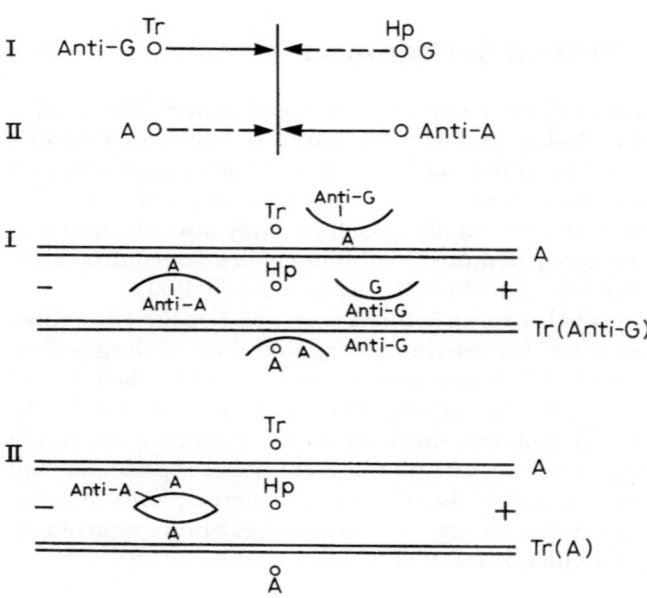

Fig. 1.1 Reaction between *Helix pomatia* (Hp) crude galactan (G) and impure anti-A from the albumin gland and *Tridacna maxima* hemolymph (Tr)
Above: I The precipitin reaction is due either to an antigalactan in Tr or to a blood group A substance in Tr reacting with anti-A from Hp (II)
Below: I If Tr has an A substance, it should not react with A substance, but with anti-A from Hp (upper trough). Unfortunately, the anti-G does cross-react with blood group A substance (peptone). The anti-A from Hp reacts with the A-substance, whereas (lower trough) the anti-G from Tr reacts with the Hp galactan and cross-reacts with A substance, both are differing in mobility.
II If Tr would have A substance instead of an anti-G, this picture here should have developed, which was not the case. The correct interpretation was always I.

14

1.3   The Source of Tridacna Lectins: The Tridacna Clams

The source of the *Tridacna* lectins is the hemolymph of these clams, where it can account for up to 50 % of the hemolymph proteins. The clams are shown from various aspects in Fig. 1.2; whereas their distribution according to Yonge (1975) can be seen in Fig. 1.3. The MW of these lectins ranges from 300,000–500,000, the MW of the subunits, obtained by -S-S- cleavage, is in the order of 22,000–44,000. Combining sites are calculated to be 10–12 in this acid protein, which is also relatively labile against proteolytic degradation and easily split by matrix-bound proteases. The main amino acids of the *Tridacnin* lectins are asparagine, glutaminic acid and glycine. The carbohydrate content is about 5–10 % [(L-fucose 3 %, D-galactose 1 % and under 1 %: D-glucose, D-mannose and both acetylated hexosamines) (Baldo et al. 1978b)].

The hemolymph components of the various Tridacnid clams have been characterized by a two dimensional electrophoretic technique, which demonstrated typical "hemolymph fingerprints" (Kania et al. 1980).

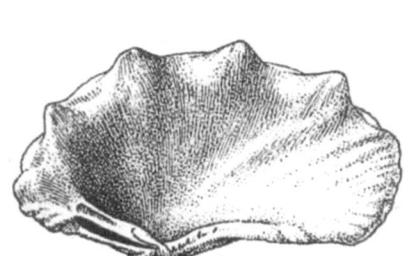 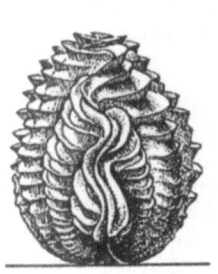 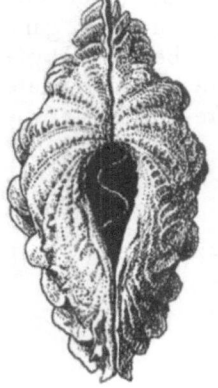

Fig. 1.2   Shells of a bivalve of the superfamily Cardiacea is shown in three views: *Tridacna maxima*, which is 5 inches long (According to Yonge 1975)

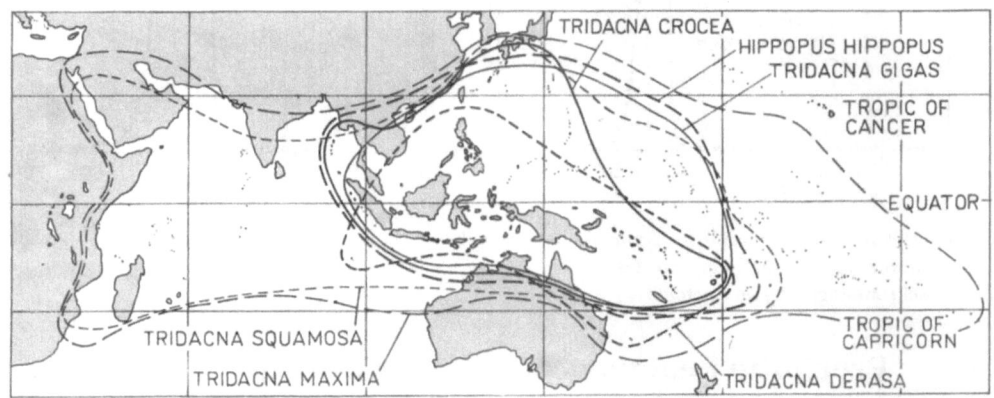

Fig. 1.3   Distribution of tridacnids shown on this map. The most widespread are *Tridacna maxima*, found from the Red Sea to Pitcairn Island, and *T. squamosa*. The most restricted are *T. gigas* and *T. crocea*. *Hippopus* and *T. derasa* are intermediate in distribution (According to Yonge 1975)

1.4   The Purification of Tridacna Lectins

Several purification methods for the *Tridacna* hemolymph lectins have been successfully applied (Baldo et al. 1978 b). Among these are such, which use affinity chromatography with a poly-arabino-galactan column (Baldo and Uhlenbruck 1975 a) or just
acid-activated Sepharose (Baldo and Uhlenbruck 1975 b). A typical example of such a
procedure is presented in Fig. 1.4.
Purification of Tridacnins was accomplished by affinity chromatography on acid-activated Sepharose 4B and on Lactogel (Medac). Starting material was the dialyzed and
freeze-dried hemolymph fluid. The main fraction of the Sepharose 4B chromatography was submitted to rechromatography on Lactogel, then dialyzed and lyophilized.
Several experiments have shown that *Tridacna* lectins can be purified by affinity
chromatography on a lactose matrix, thus demonstrating that they bind equally well
also to β-1-4-linked galactose, as already assumed by their reaction with different
asialoglycoproteins of the N-acetyl-lactosamine type.
On the other hand, the reverse method, mainly the isolation of galactan polysaccharides and galactoglycoconjugates by a Tridacnin-Sepharose 4B affinity column,
has been used by Gleeson et al. (1979). As the lectin-binding is calcium ion-dependent,
the glycoconjugates can be eluted in one step by washing the column with a calcium-
free buffer.

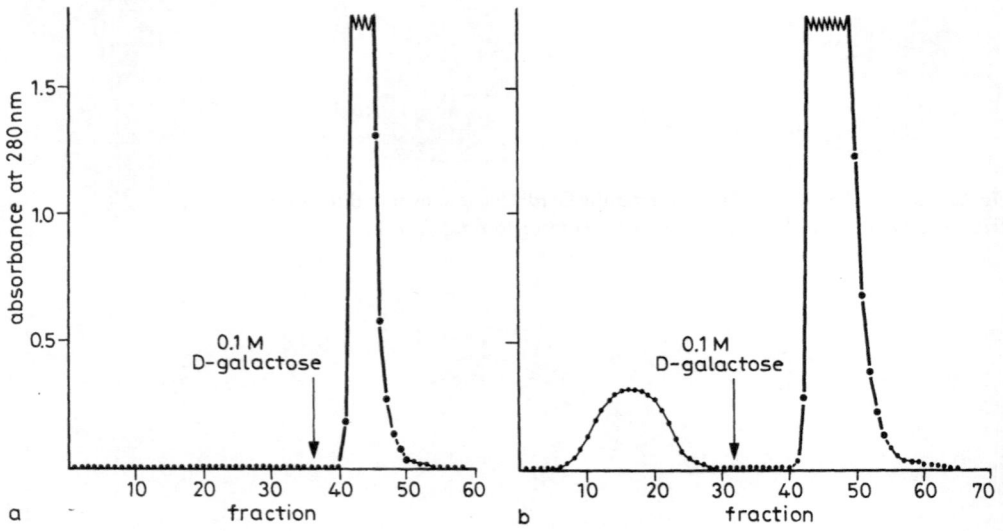

Fig. 1.4   Affinity chromatography of Tridacnin hemolymph
Above: chromatography on Sepharose 4B
Below: rechromatography on Lactogel

1.5   Precipitin Reactions of Tridacna Lectins

It was soon discovered that the *Tridacna* lectins react preferably with galactoglycoconjugates, especially galactan polysaccharides. Therefore, they have also been named antigalactans (Uhlenbruck et al. 1975 c, 1976 a). Subsequently, because of their excellent

16

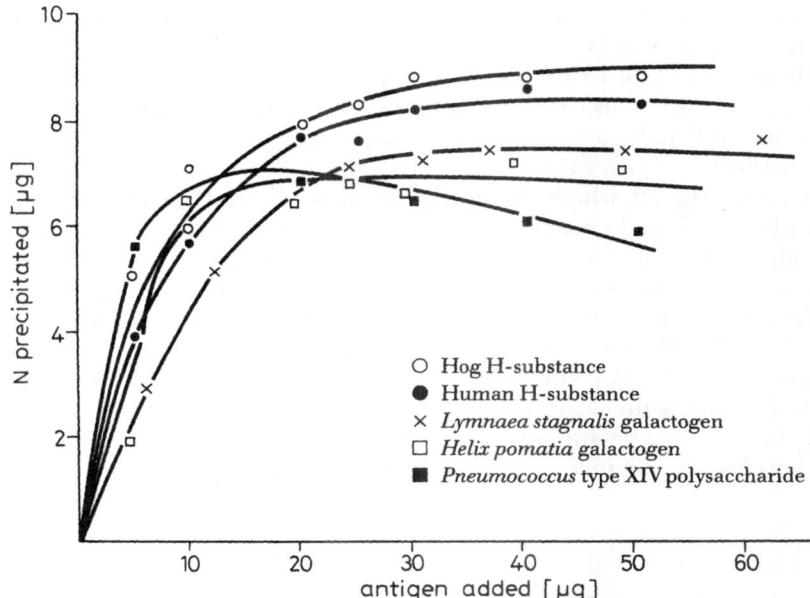

Fig. 1.5 Precipitation of some glycopeptides and polysaccharides by a *Tridacna maxima* extract. *T. maxima* extract (25 µl, 9.8 µg N) was added to each tube; total volume was 135 µl (Uhlenbruck et al. 1975 a)

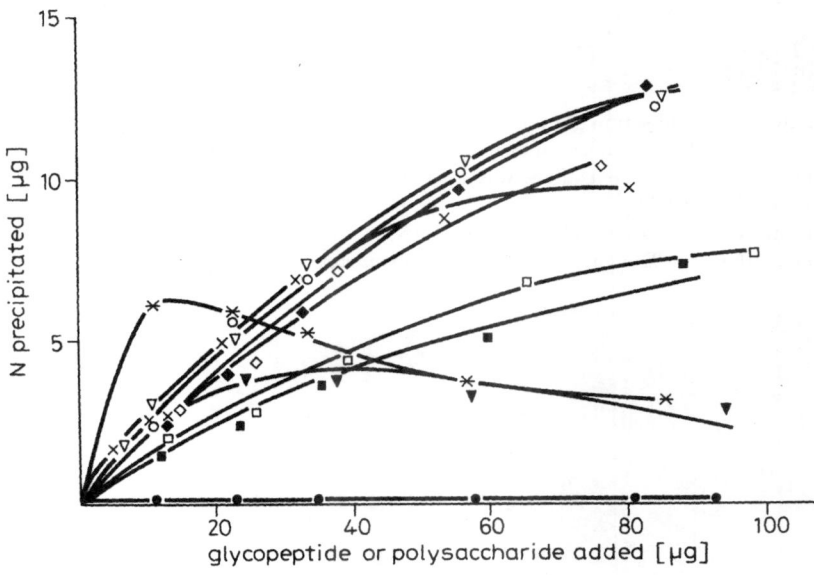

Fig. 1.6 Precipitin curves of various galactans and glycoproteins as a result of precipitation with the *Tridacna maxima* lectin

□ Larch galactan	▼ *Torulopsis groppengresseri*	◆ Bovine erythrocyte mucoid
■ Bovine lung galactan	galactomannan	● Wheat arabinogalactan-peptide
* Locust bean gum	○ *Helix pomatia* galactan	× Wheat arabinose "free" galactan
▽ *Lymnaea stagnalis* galactan	◇ Pig amnion mucoid	

17

precipitin properties, the *Tridacna* lectins have been successfully used in order to discriminate between the various galactans from different sources (Baldo and Uhlenbruck 1975c, d, Uhlenbruck et al. 1976b, 1977b, 1978b). In addition, they also "cross-react" with blood group substances (hog and human) with erythrocyte glycoproteins (bovine) and with bacterial polysaccharides (*Pneumococcus* type 14 polysaccharide) (Uhlenbruck et al. 1975a).

This is demonstrated in Fig. 1.5, where the precipitin curve of snail galactan is very similar to those of various other glycoconjugates. Figure 1.6 underlines the great variety of cross-reacting glycoconjugates in the quantitative precipitin test. The precipitin reaction of *Tridacna* lectins is especially useful for identifying asialoglycoproteins (Fig. 1.7). Some serum glycoproteins, treated with neuraminidase, are given as an example in Fig. 1.7. In Fig. 1.8 one can observe that the reverse precipitin reaction, namely, the location of a *Tridacna* lectin by its precipitin reaction with various galactans, is suitable for identifying the lectin with agar gel electrophoresis.

Anti-galactans of the *Tridacna* lectin type discriminate very well between different galactan polysaccharides, for instance in an extract of the *Achatina fulica* snail gland

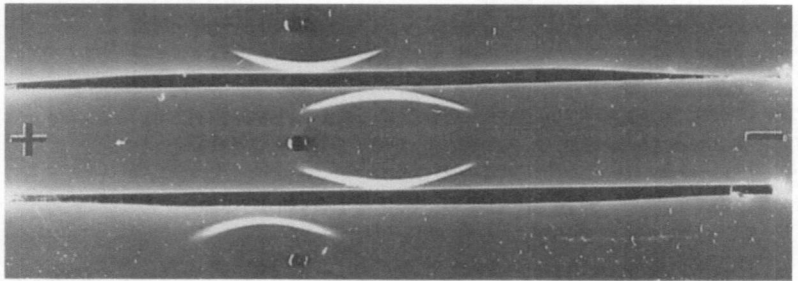

Fig. 1.7 The neuraminidase-treated serum glycoproteins were submitted to agar gel electrophoresis and made visible by their precipitin reaction with Tridacnin (trough). In all cases before neuraminidase treatment, there was no visible reaction between *Tridacna* and these serum glycoproteins
Upper well: Haemopexin (N)   Lower well: Haptoglobin (N)
Middle: β_2-Glycoprotein I (N)   Trough: Tridacnin (*Tridacna maxima*)

Fig. 1.8 Electrophoresis of *T. maxima* extract in 1% Difco Special Noble agar pH 8.6 followed by diffusion against larch arabinogalactan; *H. pomatia* was placed in the troughs after electrophoresis. Clear anode-migrating precipitin arcs are visible
Well: *T. maxima* extract, 610 µg N/ml
Top trough: arabinogalactan (SERVA), 1 mg/ml
Bottom trough: *Helix pomatia* galactan, 1 mg/ml

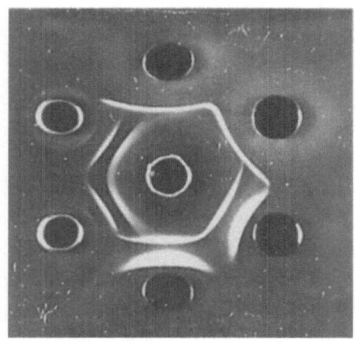

Fig. 1.9 Comparatative immunodiffusion patterns of different anti-galactans with *Achatina fulica* galactan extract
Middle well: *Achatina fulica* extract
Peripheral wells: 1.2 *Tridacna maxima*;
3.4 *Tridacna gigas*;
5.6 *Axinella polypoides*, sponge lectin

(see Fig. 1.9). Note the similar reaction type of the sponge lectin from *Axinella polypoides*. The excellent ability of the *Tridacna* lectins for precipitin reactions offers a very simple method in the search of appropriate receptors just be using the immunodiffusion technique.

When one summarizes all the glycoconjugates which have been shown to precipitate with *Tridacna* lectins, here mainly *Tridacna maxima* as a prototype of a "Tridacnin", one can obtain a list similar to Table 1.1, were all the receptor molecules for the Tridacnin from *Tridacna maxima* are given in a certain order, which implicates also certain agglutinin reactions, on which we will focus our attention in the next section. It is interesting that together with other D-galactose-specific lectins, which will be discussed later, and by using a myeloma anti-galactan from mouse, one can distinguish various plant galactans from different sources. This is documented in Table 1.2, adapted from the work of Baldo (1977).

Table 1.1 Glycosubstances which react with Tridacnin *(T. maxima)*

I Glycoproteins (purified)	II Membrane-integrated glycosubstances from/in
1. Genuine glycoproteins: a) hog gastric peptone A substance b) seminal plasma glycoprotein c) human secreted blood group substances (ABH/Lewis) d) certain serum glycoproteins e) carcinoembryonic antigen f) house dust mite allergen	a) erythrocytes d) thrombocytes b) lymphocytes e) tumor cells c) milk fat globules f) spermatozoa
	III Glycolipids
	a) neuraminic acid-free gangliosides b) glycolipids with terminal D-galactose
2. Glycoproteins after partial acid hydrolysis a) bovine erythrocyte mucoid b) secreted human blood group substances	**IV Polysaccharides**
3. Glycoproteins after neuraminidase treatment (Serum glycoproteins) a) strong reacting group b) weak reacting group	1. Galactans (or proteogalactans) a) snail galactans b) plant and alga galactans c) pneumogalactans 2. Cross-reacting polysaccharides from bacteria a) Lipopolysaccharides (*Salmonella* Perth) b) *Pneumococcus* type XIV polysaccharide c) Streptococcal B polysaccharides

19

Table 1.2 Reaction of some galactose-binding macromolecules with ara-binogalactans and galactans from wheat and ryegrass[a]
(According to Baldo 1977)

Galactose-binding preparation from	Wheat arabinogalactan	Wheat galactan	Ryegrass arabinogalactan	Ryegrass galactan
Tridacna maxima	−	+	+	+
Axinella polypoides	+ (w)	+	+	+
Cerianthus membranaceus	+ (w)	+	+	+
Abrus precatorius	−	+	+	+
RCA$_I$[b]	−	−	N.T.	−
RCA$_{II}$[b]	−	+	N.T.	+
Mouse myeloma anti-galactan J539	+	+	+	+

[a] −, No precipitation; +, precipitation; w, weak; N.T., not tested
[b] RCA$_I$ and RCA$_{II}$. *Ricinus communis* agglutinins eluted from Sepharose and separated on Sephadex G 100.

1.6   Practical Use of the Precipitin Reactions

Those lectins from invertebrates are not only of importance for the elucidation of plant arabinogalactans (Albersheim 1975; Gleeson 1977; Gleeson et al. 1978; Gleeson and Clarke 1980; Fincher et al. 1983), but also for the localization of arabinogalactan-protein structures in intracellular vesicles (Baldo et al. 1987 a). Of more practical interest is the isolation and characterization of allergens, for instance from house dust mite extracts with the help of *Tridacna maxima* lectin (Baldo and Uhlenbruck 1977), which can then be measured by a new radioallergosorbent test (Baldo et al. 1976).

Besides comparative precipitin studies with various blood group substances, polysaccharides from invertebrates, plants and bacteria (Baldo and Uhlenbruck 1975 d), the *Tridacna* lectin react preferably with serum glycoproteins, after they have been treated with neuraminidase (Uhlenbruck et al. 1977 c, d, 1978 c), a fact, which can also be seen in relationship to other lectins, for instance of plant origin (Uhlenbruck et al. 1979; Chatterjee et al. 1979).

One of the main research fields of *Tridacna* lectins is the investigation of the structural components of naturally occurring galactans, as this has been shown for the *Achatina fulica* snail galactan (Okotore and Uhlenbruck 1982 a, b), which can also be localized histochemically by fluorescent labeled *Tridacna* lectin (Okotore et al. 1982). On the other hand, galactans can be used for the isolation and characterization of *Tridacna* lectins (Janssen and Uhlenbruck 1981). Galactanases, enzymes which degrade galactans, as well as the chemical degradation of galactans (Bretting and Jacobs 1987) completely abolish the precipitin reactions with the *Tridacna* lectins.

1.7 Agglutination Patterns of Tridacna Lectins

When hemolymph samples from different Tridacnid clams were tested for their hemagglutination abilities, certain differences could be observed between the various lectins on the one hand, and different hemagglutinins, on the other (Table 1.3). The *Tridacna* lectins were tested with normal red cells and enzyme-treated erythrocytes from miscellaneous sources (Uhlenbruck et al. 1978 a).

These differences may be better understood by using Fig. 1.10 as additional help in interpreting the results, since the charge of these cells may influence the reaction or the amount of more superficially located glycoproteins and their topochemical relationship to the deeper glycolipid structures. In Fig. 1.10 a, we determined that the relationship of glycoprotein-bound neuraminic acid also reflects the relationship of membrane glycoproteins to membrane glycolipids (Fig. 1.10 a). Similar, but much more characteristic, are the changes in the electrophoretic mobility after enzyme treatment (different proteases, neuraminidase) as a reflection of the altered zeta potential. This indeed may also influence the agglutination reactions (Fig. 1.10 a, b). This is of importance, because one must be aware of the fact that besides the agglutinating lectins, so-called incomplete, nonagglutinating, but receptor-blocking lectins exist, which

Table 1.3 Hemagglutination of different erythrocytes by the hemolymph from different tridacnids

Erythrocytes	Tridacna maxima	Tridacna crocea	Tridacna gigas	Tridacna squamosa	Tridacna derasa	Hippopus hippopus
Human[a]	32	64	512	8	8	16
Human (Pro)[b]	64	128	512	32	16	32
Human (N)[c]	64	256	512	32	16	16
Pig	64	256	512	32	8	32
Pig (Pro)	512	2000	4000	64	32	128
Pig (N)	128	1000	2000	16	8	32
Rat	0	2	32	16	8	8
Rat (Pro)	16	64	256	64	8	32
Rat (N)	64	128	256	32	8	2
Rabbit	0	16	2000	64	64	64
Rabbit (Pro)	8	64	2000	32	16	32
Rabbit (N)	8	32	2000	32	32	64
Dog	2	2	128	16	8	4
Dog (Pro)	8	64	32	32	32	32
Dog (N)	64	128	512	64	16	128
Chicken	32	16	256	8	4	4
Chicken (Pro)	16	16	4000	16	4	64
Chicken (N)	128	128	1000	64	8	64
Bovine	0	0	8	4	8	8
Bovine (Pro)	0	16	32	16	16	16
Bovine (N)	256	256	128	32	8	32

[a]   no blood group specific differences detectable

[b] (Pro) Pronase-treated red cells

[c] (N)   Neuraminidase-treated erythrocytes

21

Relationship of glycoprotein-NA and glycolipoid-NA in red cells

red cell type	glycoprotein-NA	glycolipid-NA
rabbit	+	−
man	+ +	+
ox	+ + +	+ +
cat	+	+ + +

a

Fig. 1.10 Factors influencing the receptors responsible for hemagglutination by lectins
a) Different amounts of glycolipid and glycoprotein associated receptors b) influence of charge c) topochemical arrangement of receptors on charge groups

Three types of red cells with respect to charge

red cells	change in electrophoretic mobility after	
	pronase-treatment	neuraminidase-treatment
man	−50 %	− 80 %
ox	0	− 70 %
cat	+30 %	− 40 %

b

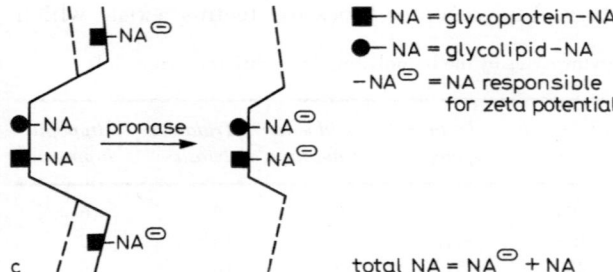

■—NA = glycoprotein−NA
●—NA = glycolipid−NA
−NA$^\ominus$ = NA responsible for zeta potential

total NA = NA$^\ominus$ + NA

c

agglutinate only after protease treatment of the cells. Sometimes, even the enzymatic removal of the neuraminic acid is sufficient, thus by reducing the charge, agglutination is possible. However, these results are difficult to interpret, because in most cases, after neuraminidase treatment *de novo* subterminal structures became available to the lectins and caused clumping of the cells. Also, the sterical hindrance of neuraminic acid has to be taken into consideration or the hindrance by glycoproteins of the availability of glycolipids.

An example of hemagglutination inhibition studies with *Tridacna* lectins is given in Table 1.4. Remarkable is that the galactans also inhibit the blood group substances H (human and hog) and that the *Pneumococcus* type 14 polysaccharide seems to be the best macromolecular inhibitor. It is interesting to note that the studies of *Tridacna* lectins were originally initiated by investigations on various anti-H reagents, and that the "H" activity of galactans was due to the presence of L-galactose (Baldo and Uhlenbruck 1974, Baldo et al. 1973, 1975).

1.8 Mitogenic Properties of Tridacna Lectins

Like all newly discovered lectins, the *Tridacna* lectins have been tested for mitogenic activity with respect to the response of human lymphocytes (Schumacher et al. 1978). The *Tridacna* lectins were found to be as stimulating to human T-lymphocytes as it is

22

Table 1.4 Results (30 min after addition of erythrocytes) of hemaggluti- nation inhibition experiments using a *Tridacna maxima* extract in con- junction with H-blood group substances, polysaccharides and invertebrate extracts (According to Uhlenbruck et al. 1975 d)

Test substance	Minimum amount of substance (μg/ml) completely inhibiting the agglutination of human group O erythrocytes by eight hemagglutinating doses of *T. maxima* extract
Human H-substance	7.2
Hog H-substance	6.6
Pig amnion mucoid 4.7 % N. S.[a]	23.4
Pig amnion mucoid 6 % N. S.[a]	24.5–49.0
Pneumococcus type XIV polysaccharide	2.1
Helix pomatia galactan	12.3
Lymnaea stagnalis galactan	9.7
Pomacea urceus galactan	2.7–5.4
Serva arabinogalactan	345.6[b]
Larch galactan	6.9–13.8[c]
Bovine pneumogalactan	22.2
Bovine red cell mucoid	125[b]
Invertebrate extracts:	
Ascaris lumbricoides	620
Littorina littorea	1500
Lumbricus rubellus	155
Lumbricus terrestris	540
Tubifex rivulorum	685

[a] N. S. – N-acetyl neuraminic acid.
[b] Trace of agglutination in all wells after 1 h.
[c] Complete inhibition at 886 μg/ml and partial inhibition from 27.7–443 μg/ml after 1 h.

known from the phytohemagglutinin PHA (Schumacher et al. 1978, Uhlenbruck et al. 1977 e, f).

The results of such experiments are shown in Fig. 1.11 and 1.12. Surprisingly (Fig. 1.11 as a control), we examined an agglutinin extract from the albumin gland and eggs of the snail *Ampullaria canaliculata*, and found a very strong mitogenic response. Pro- tease (pronase treatment) of *Tridacna* lectins abolishes the lymphocyte stimulating activity, but obviously the subunits, resulting from this treatment, are able to block not only the mitogenic action of the intact *Tridacna* lectin, but also of the closely related PHA (Fig. 1.12), with respect to the specificity. In addition we found that *Pneumococcus* type 14 polysaccharide completely inhibited the mitogenic response, a fact, which also underlines the reactivity of *Tridacna* lectins with N-acetyllactosamine structures or lactose-derived molecules. The other *Tridacna* lectins show a similarly, expressed mitogenic activity, e. g., *Tridacna gigas* (Fig. 1.11).

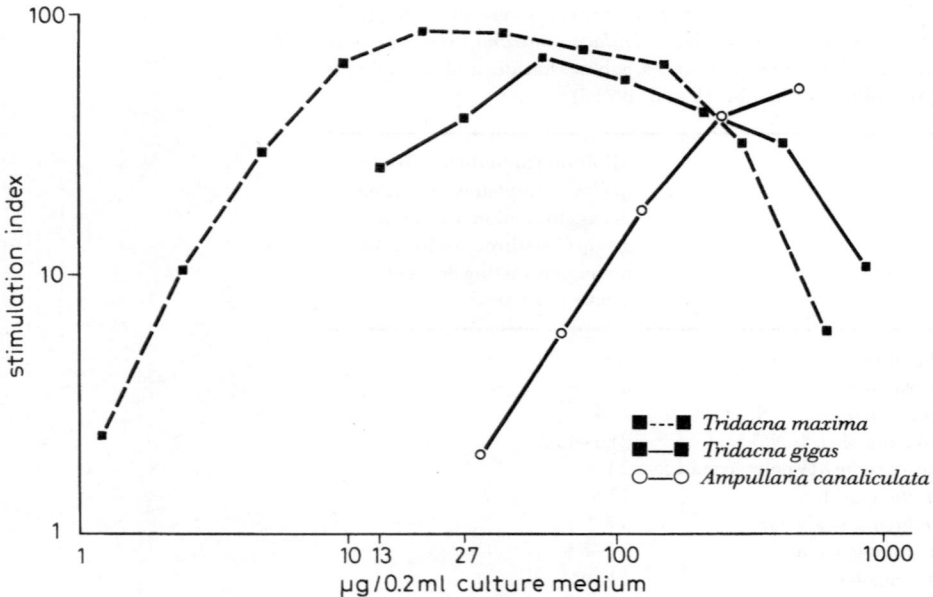

Fig. 1.11 Dose response curves of ^{3}H-thymidine uptake of lymphocytes after incubation with mitogens from *Tridacna maxima*, *Tridacna gigas* and *Ampullaria canaliculata* (prosobranch snail). These may be due to the different degrees of purification of the 3 mitogens with differing mitogenic activity (According to Schumacher et al. 1978)

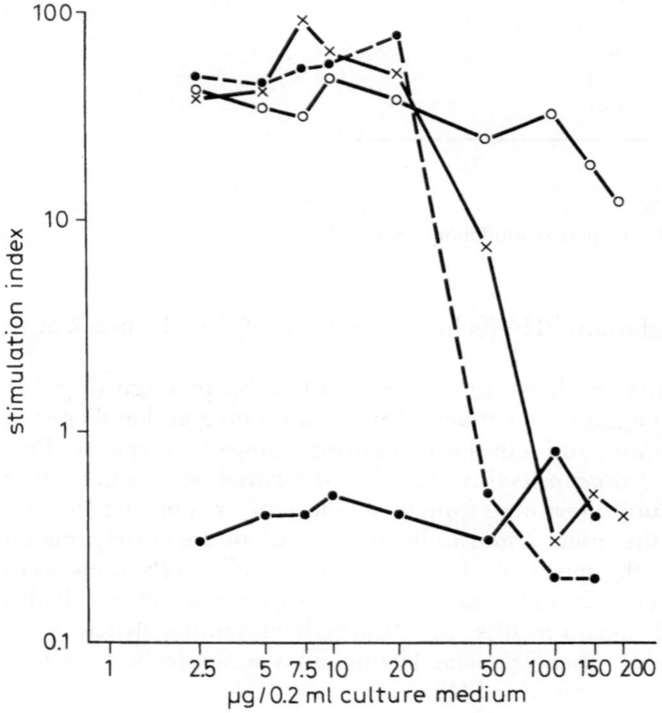

Fig. 1.12 Demonstration of the effect of pronase on the mitogenic activity of Tridacnin and of the inhibitory effect of the pronase treated fragments of Tridacnin on the mitogenic activity of Tridacnin and phytohemagglutinin

Pronase treatment of Tridacnin destroyed the mitogenic activity totally. The fragments of Tridacnin obtained by pronase treatment (*Tridacna max.*/pronase) had a significant inhibitory effect, specially on the mitogenic activity of Tridacnin in the high dose range phytohemagglutinin PHA (According to Schumacher et al., 1978)

O—O *Tridacna max.*
●—● *Tridacna max.*/pronase
×—× *Tridacna max.*/pronase
        + PHA
●---● *Tridacna max.*/pronase
        + *Tridacna max.*

1.9 The Receptor Site of Tridacna Lectins (Lectinotop)[1]

The mitogenic property of *Tridacna* lectins also raises the question: where is the receptor site or lectinotop of *Tridacna* lectins? It is generally accepted, deduced from the strong reaction with various galactans, and as has been shown especially for the *Tridacna maxima* lectin, that this anti-galactan lectin recognizes a digalactose structure, in which both sugars are linked β-1-6 glycosidically with each other (Eichmann et al. 1976).
The structure is shown in Fig. 1.13. One of the best macromolecular substances serving as an inhibitor is pneumogalactan, which carries this carbohydrate structure in the form of a side chain (Fig. 1.14).

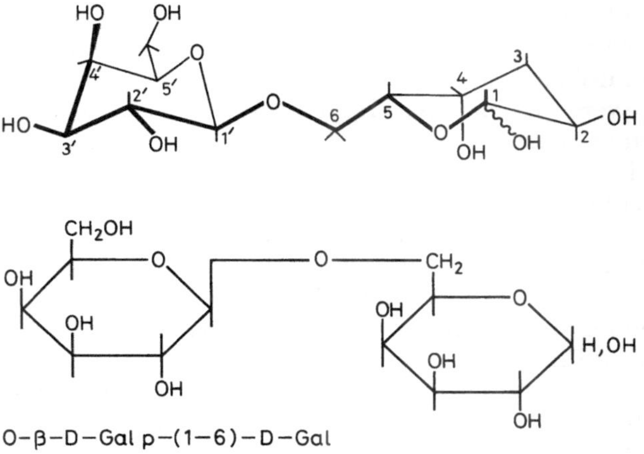
HO
OH
O
HO
OH
OH
OH
OH
CH2OH
O
CH2
H,OH
O-β-D-Gal p-(1-6)-D-Gal

Fig. 1.13 The structure of the *Tridacna maxima* receptor (Recinotop) Above: The postulated confirmation, in which the heavy bonds indicate the area of strongest binding to the immunoglobulin A J539 (and similarly *Tridacna maxima*). Adapted from Glaudemans et al. (1975)

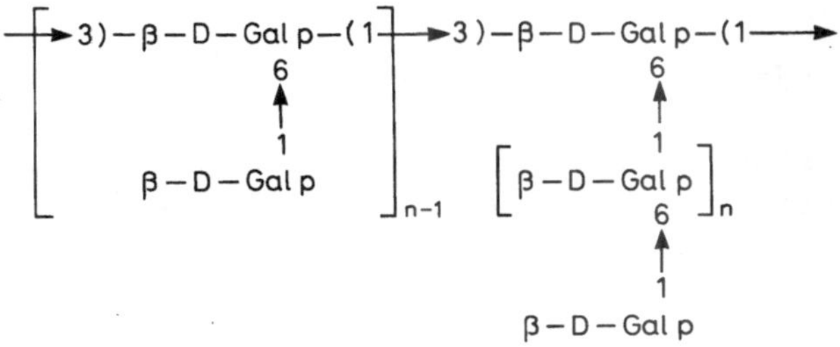
3)-β-D-Gal p-(1
3)-β-D-Gal p-(1
β-D-Gal p
β-D-Gal p
β-D-Gal p
n-1
n

Fig. 1.14 The structure of pneumogalactan from bovine lung (According to Roy and Glaudemans 1978)

Further evidence for this receptor structure has been obtained from lectinological studies with group B streptococci, since type II of these bacteria has a β-1-6-D-galactosyl epitope, which reacts with the specific classifying antisera. It was therefore logical to also check various anti-galactans as to whether they can specifically detect B streptococci of type II. In fact, this could be demonstrated (Uhlenbruck et al. 1985; Helmbold et al. 1985) also by using other antigalactan lectins, which will be considered in a later section.

[1] It has been suggested by us also, to stick to the expression "carbohydrate receptor" for the sugar structure, which is recognized by the sugar-combining site of the lectin, for which we have proposed the name "lectinotop". One could also name the "carbohydrate receptor" as "sugarotop" – suggestions, which are herewith open for discussion and consideration.

The observed cross-reactivity between anti-galactan and type II group B streptococcus antisera follows the classical definition of a cross-reaction, i. e., it is due to a common immunodominant group which is bound to a different carrier in different organisms. In the case described here the common immunodominant group is the β-D-galactosido-(1-6)-R sugar residue occurring in various polysaccharides derived from galactans of invertebrate (snail albumin gland), vertebrate (bovine lung), and plant origin (arabinogalactan) on the one hand, and the type II capsular polysaccharide of group B streptococci on the other.

Both carrier polysaccharide types differ in their composition: The backbone of the bacterial polysaccharides is mainly composed of poly-lactosyl groups whereas the galactans are composed of D-galactose and/or L-arabinose in various linkages. In this context it is interesting that nearly all known antigalactans, lectins from invertebrates, myeloma proteins or (monoclonal) antibodies react with the terminal β-(1-6)-D-galactosyl residue and accordingly can be used for the identification of type II group B streptococci (see following section). Vice versa, the anti-type II group B streptococcus antisera can serve as potent anti-galactans, which usually cannot be obtained by immunization with the poorly immunogenic "pure" galactan polysaccharides.

The "second specificity" of Tridacnins, the N-acetyl-lactosamine reactivity, may be either due to a cross-reactivity of the anti-galactan combining area or to subunits of the main lectin, which recognize a different lectinotop with a β-1-4 linked D-galactose in terminal position. However: a separation of such subunits with different lectin or carbohydrate binding properties could not be achieved so far (Uhlenbruck et al. 1975 d). Also the presence of 2 independent lectins with two different specificities could no be established. Remarkable is, that the N-acetyl-lactosamine specificity seems to be responsible for the preferred histochemical reactivity of Tridacna lectins with various human tumor cells (unpublished results).

1.10 Cross-Reacting Lectins and Antibodies

Cross-reacting lectins and antibodies have to be mentioned in this connection, because they also react with the *Tridacna maxima* specific receptor, namely the β-D-Gal(1-6)-D-Gal disaccharide structure. This has been convincingly demonstrated by Eichmann et al. (1976). According to these fundamental experiments one can divide the β(1-6)-D-Gal antigalactans into three groups:

1. The *Tridacna* lectins, which have also been thoroughly discussed in previous reviews (Uhlenbruck et al. 1974, 1975 d).

2. Sponge lectins, e. g., from *Axinella polypoides* (Eichmann et al. 1976, Bretting et al. 1983) or from the sea anemone *Cerianthus membranaceus* (Baldo et al. 1977 b), which have the same or a very similar "combining site".

3. The third group is represented by the immunoglobulin class of anti-galactans, which can be subdivided into three groups:

    a) The myeloma proteins (mouse), a "classical" model of an anti-galactan (Glaudemanns 1975). This group has been exhaustively studied with respect to

(1) their specificity (Glaudemans et al. 1975, Manjula and Glaudemans 1976), (2) their reaction with house dust mite extracts (Baldo et al. 1977 a), (3) in a comparison of their idiotypes compared with antigalactan antibodies from mice (Mushinski and Potter 1977), (4) their reaction with artificial antigens (Ittah and Glaudemans 1981; Bhattacharjee et al. 1981), (5) models for their specific space-filling grooves and cavities (Feldman et al. 1981), or surface type (Ekborg et al. 1983, Glaudemans et al. 1984), (6) and by probing the combining site of these anti-β(1-6)-D-galactopyranans (Glaudemans and Kovac 1985) which showed that the antibody binds to interantenary galactosyl groups of the antigen (Glaudemans et al. 1986). For review see Rudikoff et al. 1983, a paper in which also the diversity and structure of the respective idiotypes is discussed, a subject which has recently been reanalyzed with respect to genetic aspects by Rudikoff (1988). Their usefulness for the investigation of plant arabinogalactans has been elucidated by Baldo et al. (1978 a).

b) A second group is composed of a large number of monoclonal antibodies, which have the same "anti-galactan" specificity (Glaudemans 1987; Rudikoff 1988) as has been proved by immunization with *Helix pomatia* snail galactan (Sölter et al. 1985). Those antibodies can be successfully used for the histochemical topobiology of galactans (Okotore et al. 1982; Dienst et al. 1986). In this group one can probably also integrate certain antibacterial antibodies with anti-galactan specificity for instance against group B streptococcus type II (Uhlenbruck et al. 1985).

c) A third classification implicates the inclusion of the snake lectins, as a representative example of vertebrate lectins, which have been shown to have a very similar β-D-(1-6)-galactopyranan combining site (Uhlenbruck and Helmbold 1985).

d) A fourth group may be represented by Calcium-activated neutral lectin-like proteases (Calpains), which bind to the same lectinotop (Zimmermann and Schlaepfer 1988).

Those considerations are mainly of theoretical interest and lead to the question of cross-reactivity of the various anti-idiotype antibodies between anti-galactans of the myeloma or immunoglobulin type and those of the lectin type. The anti-galactan reactivity of C-reactive protein-like anti-galactans (Uhlenbruck et al. 1982) could not only be attributed to a lectin-like specificity, but mainly to a reaction with phosphate groups (Sölter and Uhlenbruck 1986).

1.11   The Biological Role of Tridacna Lectins

The biological role of *Tridacna* lectins has been extensively discussed in relationship to the so-called invertebrate immunity (Hildemann and Benedict 1975): For the *Tridacna* lectins we assumed that they contribute less to a defense mechanism, but more to symbiosis with algae, which they culture and house for their nutrition (Uhlenbruck and Steinhausen 1977) by using the "old" algae for their metabolism. The lectins may serve to eliminate galactan-containing membrane constituents or to stimulate the growth of the algal symbionts. This would fit into the concept of algal symbiosis (Taylor 1973; Muscatine and Greene 1973) as well as into consideration on molluscan immunobiology (Bayne 1983).

With respect to the defense mechanism it is remarkable to note that most microorganisms have α or $\beta(1-4)$-linked galactose residues, mostly in the furanose form (Bardalaye and Nordin 1976; Bennett et al. 1985), in their mixed galactans, and that it is rather the exception when a mycoplasma galactan with a 6-0-β-D-galactopyranosyl-D-galactose unit cross-reacts with a bovine lung galactan, a fact, which may inhibit the immune reaction of the host (Kakoma and Kinyanjui 1974; Shifrine and Gourlay 1965).

In general, it can be concluded that the biological role of the *Tridanca* lectins is not completely established and has to be further investigated beyond any speculation. Besides being scientific tools for the structural studies on various galactans, they have also proved to be excellent tools in serological and chemical methods (Bretting et al. 1981). So far, the various anti-galactans from different sources have enriched the field of immunochemistry and immunobiology in most interesting and stimulating aspects and have integrated lectinology into immunology as an alternative system of recognition.

1.12   References

Albersheim P (1975) The walls of growing plant cells. Sci Am vol. 232 April 1975: 80–105

Baldo BA (1977) Galactose-binding macromolecules. Arabinogalactan Proteins News. agp University of Melbourne Press 1:21

Baldo BA, Uhlenbruck G (1973) Cross-reactive human blood group H-active polysaccharide from *Helix pomatia*. I. Detection with catfish anti-H and eel sera. Immunology 25:649–661

Baldo BA, Uhlenbruck G (1974) Studies on the agglutinin specificities and blood group 0(H)-like activities in extracts from the molluscs *Pomacea paludosa* and *Pomacea urceus*. Vox sang 27:67–80

Baldo BA, Uhlenbruck G (1975a) Tridacnin, a potent antigalactan precipitin from the hemolymph of *Tridacna maxima* (Röding) In: Hildemann WH and Benedict AA (eds) Immunologic phylogeny. Advances in experimental medicine and biology 64. Plenum Press. New York London pp 3–11

Baldo BA, Uhlenbruck G (1975b) Purification of Tridacnin, a novel anti-β-(1-6)digalactobiose precipitin from the haemolymph of *Tridacna maxima* (Röding). FEBS Lett 55:25–29

Baldo BA, Uhlenbruck G (1975c) Anti-galactan activity in *Tridacna maxima* (Röding) haemolymph. Immunology 29:1161–1170

Baldo BA, Uhlenbruck G (1975d) Quantitative precipitin studies on the specificity of an extract from *Tridacna maxima* (Röding). Carbohydr Res 40:143–151

Baldo BA, Uhlenbruck G (1977) Selective isolation of allergens. Clin Allergy 7:429–443

Baldo BA, Uhlenbruck G, Steinhausen G (1973) Blood group 0(H)-like activity in extracts from some invertebrates. Vox sang 25:398–410

Baldo BA, Uhlenbruck G, Salfner B (1975) Studies on the specificities of various anti-H reagents. Z Immunitätsforsch 148:330–340

Baldo BA, Turner KJ, Uhlenbruck G (1976) Isolation of a potent allergen from house dust mite by interaction with the lectin Tridacnin. Experientia 32:641–643

Baldo BA, Fletcher TC, Uhlenbruck G (1977a) Reaction of house dust mite extracts with mouse anti-phosphorylcholine and antigalactan myeloma proteins. Naturwissenschaften 64:594

Baldo BA, Uhlenbruck G, Steinhausen G (1977b) Invertebrate anti-galactans. A comparative study of agglutinins from the clam *Tridacna maxima*, the marine sponge *Axinella polypoides* and the anemone *Cerianthus membranaceus*. Comp Biochem Physiol 56A:343–351

Baldo BA, Neukom H, Stone BA, Uhlenbruck G (1978a) Reaction of some invertebrate and plant agglutinins and a mouse myeloma anti-galactan protein with an arabinogalactan from wheat. Aust J Biol Sci 31:149–160

Baldo BA, Sawyer WH, Stick RV, Uhlenbruck G (1978b) Purification and characterization of a galactan-reactive agglutinin from the clam *Tridacna maxima* (Röding) and a study of its combining site. Biochem J 175:467–477

Bardalaye PC, Nordin JH (1976) Galactosaminogalactan from cell walls of *Aspergillus niger*. J Bacteriol 125 (No 2) 655–669

Bayne CJ (1983) Molluscan immunobiology. The Molluscan 5:407–486 (Physiology, Part 2)

Bennett JE, Bhattacharjee AK, Glaudemans CPJ (1985) Galactofuranosyl groups in *Aspergillus fumigatus* galactomannan. Mol Immunol 22:224–251

Bhattacharjee AK, Das DK, Roy A, Glaudemans CPJ (1981) The binding sites of the two monoclonal immunoglobulin AJ539 and W3129. Thermodynamic mapping of a groove- and a cavity-Type immunoglobulin, both having antipolysaccharide specificity. Mol Immunol 18:277–280

Bretting H, Jacobs G (1987) The reactivity of galactose oxidase with snail galactans, galactosides and D-galactose-composed oligosaccharides. Biochim Biophys Acta 913:342–348

Bretting H, Whittaker NF, Kabat EA, Königsmann-Lange K, Thiem HJ (1981) Chemical and immunochemical studies on the structure of four snail galactans. Carbohydr Res 98:213–236

Bretting H, Jacobs G, Dannenberg F (1983) Structural similarities between *Helix Pomatia* galactan and the carbohydrate moiety of hemocyanin. Comp Biochem Physiol 75B:269–276

Chatterjee BP, Vaith P, Chatterjee S, Karduck D, Uhlenbruck G (1979) Comparative studies of new marker lectins for alkali-labile and alkali-stable carbohydrate chains in glycoproteins. Int J Biochem 10:321–327

Dienst C, Böhmer G, Uhlenbruck G (1986) Monoklonale Antikörper und ihre Anwendung in der Pulmonologie. Atemwegs-Lungenkr 12:138–139

Eichmann K, Uhlenbruck G, Baldo BA (1976) Similar combining specificities of invertebrate precipitins and mouse myeloma protein J 539 for β-(1-6)-galactans. Immunochemistry 13:1–6

Ekborg G, Ittay Y, Glaudemans CPJ (1983) Monoclonal IgA J539 binds galactosylpyranosyl antigens on its surface. Mol Immunol 20:235–238

Feldmann RJ, Potter M, Glaudemans CPJ (1981) A hypothetical space-filling model of the V-regions of the galactan-binding myeloma immunoglobulin J539. Mol Immunol 18:683–698

Fincher GB, Stone BA, Clarke AE (1983) Arabinogalactan-proteins: structure, biosynthesis, and function. Ann Rev Plant Physiol 34:47–70

Glaudemans CPJ (1975) The interaction of homogeneous, murine myeloma immunoglobulins with polysaccharide antigens. Adv Carbohydr Chem Biochem 31:313–346

Glaudemans, CPJ (1987) Seven structurally different murine monoclonal galactan-specific antibodies show identity in their galactosyl-binding subsite arrangements. Mol Immunol 24:371–377

Glaudemans CPJ, Kovac P (1985) Probing the combining site of monoclonal IgA J539 using deoxyfluoro- and other galactosides as ligands. Mol Immunol 22:651–653

Glaudemans CPJ, Zissis E, Jolley ME (1975) Binding studies on a mouse-myeloma immunoglobulin A having specificity for β-D-(1-6)-linked D-galactopyranosyl residues. Carbohydr Res 40:129–135

Glaudemans CPJ, Kovac P, Rasmussen K (1984) Mapping of subsites in the combining area of monoclonal anti-galactan immunoglobulin A J539. Biochemistry 23:6732–6736

Glaudemans CPJ, Bhattacharjee AK, Manjula BN (1986) Monoclonal anti-galactan IgA J 539 binds intercatenarily to its polysaccharide antigen. Observations on the binding of antibody to a macromolecular antigen. Mol Immunol 23:655–660

29

Gleeson PA (1977) Isolation of arabinogalactan proteins using insolubilized galactose-binding macromolecules. In: Arabinogalactan protein news. Proc Arabinogalactan Protein Club 1:30–33

Gleeson PA, Clarke AE (1980) Antigenic determinants of a plant proteoglycan, the *Gladiolus* style arabinogalactan-protein. Biochem J 191:437–447

Gleeson PA, Clarke AE, Jermyn MA, Knox RB (1978) Arabinogalactan proteins of the female reproductive tissues of *Gladiolus gandavensis*. Proc Aust Bio Soc II p 33

Gleeson PA, Jermyn MA, Clarke AE (1979) Isolation of an arabinogalactan protein by lectin affinity chromatography on Tridacnin-Sepharose 4B. Anal Biochem 92:41–45

Griffiss J McL, Goroff DK (1981) Immunological cross-reaction between a naturally occurring galactan, agarose, and an LPS locus for immune lysis of *Neisseria meningitides* by human sera. Clin Exp Immunol 43:20–27

Helmbold W, Prokop O, Uhlenbruck G, Böhmer G, Lütticken R (1985) Snake venom lectins: a new group of T- and B-cell mitogenic anti-galactans. Biomed Biochim Acta 44:K91–96

Hildemann WH, Benedict AA (eds) (1975) Immunologic Phylogeny. Advances in experimental medicine and biology 64. Plenum Press, New York London

Ittah Y, Glaudemans CPJ (1981) Preparation of two methyl desoxyfluoro-β-D-galactopyranoside, and their interaction with galactan-specific immunoglobulin A J539 (FAB'). Carbohydr Res 95:189–194

Janssen E, Uhlenbruck G (1981) Purification of Tridacnin and related galactophilic lectins. 3rd Symp Lectins in Cell Biology and Medicine, Nov 20–21, Hamburg, (Abstr) p 7

Kakoma I, Kinyanjui M (1974) Immunogenicity of mycoplasma galactan. Res Vet Sci 17:397–399

Kania J, Janssen E, Uhlenbruck G, Pearson R (1980) Comparative investigations on the protein components of the haemolymph fluid of the tridacnid clam family using two-dimensional gel electrophoresis. Comp Biochem Physiol 66B:117–121

Manjula BN, Glaudemans CPJ (1976) Homogeneous, anti-galactan immunoglobulins. The questions of specificity. Immunochemistry 13:469–471

May F, Weinbrenner H (1938) Über die Ablagerung des Galaktogens im Warmblüterorganismus. Z Biol 99:199–216

Muscatine L, Greene RW (1973) Chloroplasts and algae as symbionts in molluscs. In: Int Rev Cytology. Academic Press, New York London. 36:137–169

Mushinski E, Potter M (1977) Idiotypes on galactan binding myeloma proteins and anti-galactan antibodies in mice. J Immunol 119:1888–1893

Okotore RO, Uhlenbruck G (1982a) Lectin receptors in proteogalactans from the albumin gland of *Achatina fulica*. In: Bog-Hansen TC (ed) Lectins. Biology, Biochemistry, Clinical Biochemistry. Vol II. de Gruyter Berlin New York, pp 351–365

Okotore RO, Uhlenbruck G (1982b) Additional lectin receptors in galactans from the albumin gland of the *Achatina fulica* snail. Experientia 38:507–508

Okotore RO, Ortmann M, Karduck D, Klein PJ, Uhlenbruck G (1982) Histochemical distribution of certain biochemical constituents in the albumin glands of snails. J Histochem Cytochem 30:895–900

Prokop O, Uhlenbruck G (1970) Protektine. In: Perlick E, Plenert W, Prokop O (eds) Fortschritte der Hämatologie. Vol 1. Barth, Leipzig, pp 17–39

Renwrantz LR, Cheng TC (1977) Identification of agglutinin receptors on hemocytes of *Helix pomatia*. J Invertebr Pathol 29:88–96

Roy N, Glaudemans CPJ (1978) On the structure of mammalian-lung galactan. Carbohydr Res 63:318–322

Rudikoff S (1988) Antibodies to β(1,6)-D-galactan: proteins, idiotypes and genes. Immunol Rev 105:97–111

Rudikoff S, Pawlita M, Pumphrey J, Mushinski E, Potter M (1983) Galactan-binding antibodies. Diversity and structure of idiotypes. J Exp Med 158:1385–1400

Schumacher G, Uhlenbruck G, Chatterjee BP, Steinhausen G, Vaith P, van Mil A (1978) A new group of mitogenic lectins from invertebrate sources. Z Immunitätsforsch 154:62–74

Shifrine M, Gourlay RN (1965) Serological relationship between galactans from normal bovine lung and from *Mycoplasma mycoides*. Nature (Lond) 208:498–499

Sölter J, Uhlenbruck G (1986) The role of phosphate groups in the interaction of human C-reactive protein with galactan polysaccharides. Immunology 58:139–144

Sölter J, Uhlenbruck G, Düvel H, Raftery B, Mohr R (1985) A monoclonal antibody to the galactan from *Helix pomatia* snails recognizes β-(1-6)-linked D-galactose residues. Immunobiology 169:330–334

Taylor DL (1973) Algal symbionts of invertebrates. Ann Rev Microbiol 27:171–187

Uhlenbruck G, Helmbold W (1985) Schlangengift als Quelle von B- und T-Lymphozyten-Mitogenen. Dtsch Ärztebl 82:3176–3177

Uhlenbruck G, Steinhausen G (1977) Tridacnins: symbiosis profit or defense-purpose? Dev Comp Immunol 1:183–192

Uhlenbruck G, Sprenger I, Ishiyama I (1971) A new polyvalent proteinase-inhibitor occurring in the albumine gland of *Helix pomatia*. Z Klin Chem Klin Biochem 9:361–362

Uhlenbruck G, Sprenger I, Hermann G, Franke H (1972) Fish eggs: another new source of polyvalent proteinase-isoinhibitors. Z Naturforsch 27b:322–323

Uhlenbruck G, Dahr W, Rothe A, Baldo BA (1974) Fakten und Folgerungen aus Forschungsergebnissen von Erythrozyten-Rezeptoren, heterophilen Agglutininen und Tumorzellmembranen. Westdeutscher Verlag, D-5670 Opladen Forschungsbericht Nr. 2475

Uhlenbruck G, Baldo BA, Steinhausen G (1975a) Anti-carbohydrate precipitins and haemagglutinins in haemolymph from *Tridacna maxima* (Röding) Z Immunitätsforsch 150:354–363

Uhlenbruck G, Steinhausen G, Gauwerky Ch, Baldo BA, Renwrantz L (1975b) Über zwei weitere Präzipitine gegen das Galaktogen aus *Helix pomatia*. Biol Zentrabl 94:205–210

Uhlenbruck G, Steinhausen G, Baldo BA, Kareem HA (1975c) Antigalactans: new classes of anti-carbohydrate precipitins. Naturwissenschaften 62:301–302

Uhlenbruck G, Steinhausen G, Baldo BA (1975d) Galactane und Antigalactane. Stippak, D-5100 Aachen, Postfach 1262

Uhlenbruck G, Steinhausen G, Baldo BA (1976a) Galactans and anti-galactans from invertebrates. Z Naturforsch 31c:205–206

Uhlenbruck G, Steinhausen G, Kareem HA (1976b) Different glycosubstances and galactans in the albumin gland and eggs of *Achatina fulica*. Z Immunitätsforsch 152:220–230

Uhlenbruck G, Steinhausen G, Baldo BA (1977a) Different anti-galactans in the haemolymph of *Tridacna maxima* and *Tridacna gigas*. Comp Biochem Physiol 56B:329–333

Uhlenbruck G, Steinhausen G, Palatnik M (1977b) Similarity of glycoproteo-galactans in the albumin glands from *Achatina* and *Borus* snails. Comp Biochem Physiol 57B:335–339

Uhlenbruck G, Steinhausen G, Schwick HG (1977c) A new marker for neuraminidase-treated human serum glycoproteins from the haemolymph of *Tridacna maxima* (Röding). J Clin Chem Clin Biochem 15:21–26

Uhlenbruck G, Haupt H, Reese I, Steinhausen G (1977d) Serumcholinesterase als Modell eines Glykoproteins. J Clin Chem Clin Biochem 15:561–564

Uhlenbruck G, Schumacher K, Steinhausen G, van Mil A (1977e) Tridacnin, a new mitogenic lectin from invertebrate sources. Z Immunitätsforsch 153:265–267

Uhlenbruck G, Schumacher K, Steinhausen G, van Mil A (1977f) Über eine neue Gruppe von mitogen wirkenden Lektinen aus Invertebraten. In: Schumacher K, Grosser KD (eds) Aktuelle Probleme der Inneren Medizin. Schattauer. Stuttgart, pp 187–193

Uhlenbruck G, Karduck D, Pearson R (1979) Different Tridacnins in different tridacnid clams: a comparative study. Comp Biochem Physiol 63b:125–129

Uhlenbruck G, Steinhausen G, Geserick G, Prokop O (1978b) Further comparative studies of glycosubstances and proteins from different snail albumin glands. Comp Biochem Physiol 59b:285–288

Uhlenbruck G, Baldo BA, Steinhausen G, Schwick HG, Chatterjee BP, Horejsi V, Krajhanzl A, Kocourek J

31

(1978c) Additional precipitation reactions of lectins with human serum glycoproteins. J Clin Chem Clin Biochem 16:19–23

Uhlenbruck G, Vaith P, Karduck D, Haupt H, Müller WEG (1979) Anti-galactan Lectins. In: Peeters H (ed) Protides of Biological Fluids. Pergamon Press, Oxford New York, pp 595–598

Uhlenbruck G, Janssen E, Javeri S (1982) Two different anti-galactan lectins in eel serum. Immunobiology 163:36–47

Uhlenbruck G, Lütticken R, Böhmer G, Janssen E, Pulverer G (1985) Group B streptococcus type II antisera have antigalactan specificities. Zentralbl Bakteriol Hyg A 259:179–187

Yonge CM (1975) Giant clams. Sci Am 232:96–105

Zimmerman UJP, Schlaepfer W (1988) Calcium-activated neutral proteases (Calpains) are Carbohydrate Binding proteins. J Biol Chem 263:11609–11612

2 Mistletoe Lectins (2)
Hartmut Franz

2.1 Introduction

In 1988 I wrote in my review on Viscaceae lectins (Franz 1989): "The author of this review hopes for progress in this field over the next years. Moreover, he is anticipating it." The recent development seems to confirm this remark. A number of papers published recently might justify the presentation of a further review about mistletoe lectins.

The special interest in mistletoe lectins is due to two factors. First, mistletoe lectins are strongly related to ricin and abrin. Second, these lectins are involved at different concentrations of the therapeutically used preparations from *Viscum album*.

Topics of this review are cytotoxicity, toxicity in vivo, mistletoe lectins and cancer, mediator release in vivo and in vitro, and the preparation of immunotoxins containing the A chain of mistletoe lectin I (ML I).

2.2 Cytotoxicity

The toxic lectins are internalized into mammalian cells by receptor-mediated endocytosis. Following internalization toxin molecules are released into the cytosol and inhibit essential cellular functions, mainly protein synthesis. Of special interest are the intracellular trafficking and the cytotoxic process of toxic lectins. Yoshida et al. (1991) studied the cytotoxicity of ML I in CHO and V 79 cells and in mutant cell lines altered in Golgi functions or in endosomal acidification.

Figure 2.1 demonstrates that the cytotoxicity of ML I is enhanced by NH_4Cl (20 mM) and by 10 mM Nigericin (ionophore for monovalent cations abolishing the proton gradient by electroneutral transmembrane exchange of protons for monovalent cations). A CHO mutant defective in endosomal acidification (DMP[R]-2) which is resistant to diphtheria toxin, modeccin and *Pseudomonas aeruginosa* exotoxin A and hypersensi-

Table 2.1 The cytotoxicity of ML I and ricin in CHO, Mon[R]-31, V79, and MF-1 cells.

Cell Lines	LD_{50} of ML I	LD_{50} of Ricin
	(ng/ml)	(ng/ml)
CHO	30,000	75
Mon[R]-31	2,000	25
V79	>100	5
MF-1	20	1.5

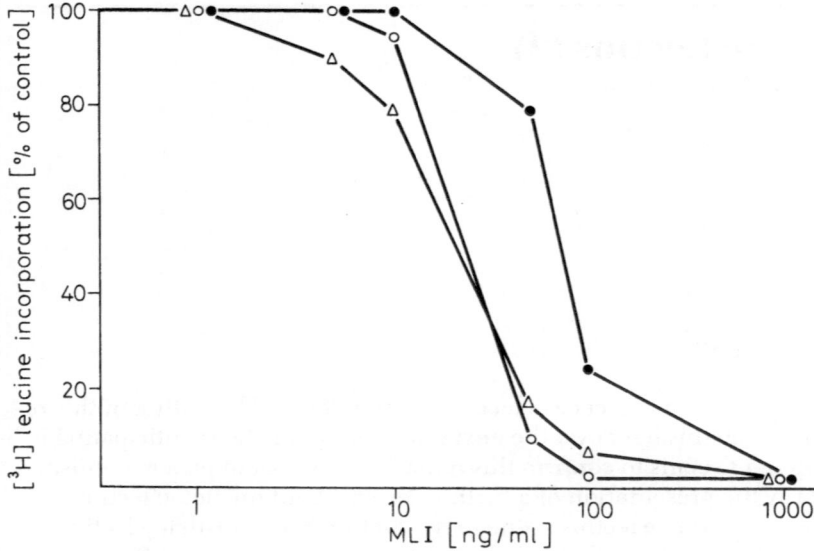

Fig. 2.1 The effect of NH₄Cl or nigericin on the cytotoxicity of ML I in Vero cells. The monolayer culture was incubated in α-MEM with or without NH$_4$Cl (20 mM) or nigericin (10 nM) for 1 h at 37 °C. The cells were exposed to different concentrations of ML-I in α-MEM containing 10 % fetal bovine serum for 5 h at 37 °C, and then incubated with (^{3}H)leucine (0.5 µCi/ml) for 45 min at 38 °C in leucine-free medium for the measurement of protein synthesis in vivo. ●, control; △, 20 mM NH$_4$Cl; ○, 10 nM nigericin

tive to ricin showed increased sensitivity to ML I. Mon^R-31 and MF-1 are monensin- and compactin-resistant mutants derived from CHO and V 79 cell lines. Both are presumably altered in Golgi functions. The cytotoxicity of ML I was found to be increased in both Mon^R-31 and MF-1 cells as compared with their parenteral cells. These results indicate that the effect of chemicals or mutations altering the endosomal acidification or Golgi functions on the cytotoxicity of ML I, is similar to those on the cytotoxicity of ricin. The functional similarity is shown in Table 2.1. The Golgi region seems to be involved in the intoxication process caused by both toxic lectins.

2.3 In Vivo Toxicity

The considerable toxicity of toxic lectins after parenteral application is not yet clear. It seems to be a rather complex phenomenon (Franz 1989). Because histological routine investigations (Holle, Zschiesche, pers. commun.) did not demonstrate cell injury at a larger scale we (Gossrau and Franz 1990) studied mice histochemically after intraperitoneal application of different doses of ML I after different time intervals. Various plasma membrane-associated hydrolases, transferases, and oxidases as well as Golgi apparatus- and endoplasmic reticulum-linked hydrolases, peroxisomal and extraperoxisomal oxidases, lysosomal hydrolases, mitochondrial dehydrogenases, the cytoskeletal proteins keratin and vimentin as well as iron, glycogen and lipids were analyzed in all organs and tissues of female mice.

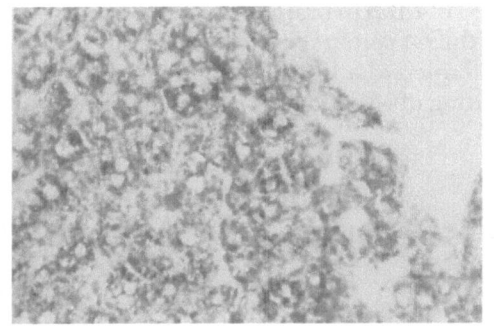

Fig. 2.2 a Demonstration of glycogen in the normal liver of mouse.
b Glycogen disappears completely after application of ML I (600 ng ML I i. p., 24 h)

Independent of the doses administered (between 300 and 1200 ng/mouse) up to now a clear cut response was found in the liver.

Glycogen disappeared completely from all hepatocytes independent of the ML I-concentration and exposure time (Fig. 2.2), whereas the activity of Golgi-associated thiamine pyrophosphatase in hepatocytes increases. The activity of nonspecific alkaline phosphatase in the sinusoidal endothelial cells depended on the amount of applied ML I and the time of treatment. Although the results concerning glycogen, thiamine phosphatase, and nonspecific alkaline phosphotase are highly reproducible, these results do not explain the death of the animals.

The response of the liver appears to be primarily nontoxic because typical signs of toxicity, e. g., lipid accumulation, reduced or absent activities of peroxisomal hydrogen peroxide-generating oxidases, disturbances of plasma membrane-associated hydrolases at the biliary pole of hepatocytes, or an increase in iron in their cytoplasm as observed after treatment of mice and/or rats with salicylate, valproate, glucocorticoids, or high concentrations of Zn (for references, see Gossrau et al. 1989 a, b, 1990) never occured following ML I administration. Especially lipid accumulation and disturbances in lipid metabolism are considered typical toxic (and early) events in hepatocytes and their injury or are even postulated for hepatocyte toxicology. Perhaps normal cells (carrying "low affinity receptors" for ML I) are protected against the cytotoxicity of ML I by complex formation by this lectin with glycoconjugates of body fluids.

According to our present understanding of the complex phenomenon for in vivo toxicity, an interaction of ML I with the CNS or the release of mediators might be responsible.

Recently Ritchie (1990) has reported the stimulation of the release of hepatic glucose from glycogen by interleukin 6 (IL-6) from human monocytes and by recombinant human IL-6. Only antisera to IL-6 were capable of reducing the glucose releasing factor activity.

These data suggest a participation of cytokines in the intoxication of animals by ML I. Hajto et al. (1889) demonstrated the release of serum IL-6 after injection of ML I (1 ng/kg) into tumor patients.

The drastic changes of glycogen and enzyme activities were not observed after administration of both the A- and B-chain of ML I alone and after injection of A- and B-chains in intervals of one hour even when using concentrations higher than that of ML I.

Recently Franz (1990b) has summarized the use of ML I in brain histochemistry. There are ML I binding sites on the synaptosomes of the rat cerebral cortex. The main advantage of ML I seems to be the demonstration of microglia first of all in the brain of rodents. It is still uninvestigated whether the binding of ML I to microglia is connected to activation and/or mediator release.

2.4   Mistletoe Lectins and Cancer Therapy

2.4.1   Release of IL-1 and IL-2

It is not the aim of this review to reflect the incongruous opinions about the effectivity of numerous commercial mistletoe preparations in cancer treatment. Here, only biological activities of mistletoe lectins which might be of therapeutical interest are taken into account.

It is not surprising that first the cytotoxic effect of ML I and the other lectins from *Viscum album* was discussed. Some authors found that ML I is much more toxic against a variety of tumor cell lines than against nonmalignant cells (for review see Franz 1989a). On the other hand, the results of Al Alousi et al. (1990) demonstrated clearly that ML I binds only to a certain portion of breast cancer cells. Most investigators agree that the cytotoxic effect of the lectins alone is not the only reason for the pharmacological interest in mistletoe lectins.

More and more it has been noted that the interaction of ML I with immunocompetent cells leads to mediator release and immunomodulation. Already in 1982 we published (together with Walzel et al.) a well reproducible foot pad swelling test in mice indicating the localized release of inflammatory mediators. Later, we reported on the release of a macrophage stimulating factor from peripheral mononuclear cells (PMNC) (Metzner et al. 1985) and the release of IL-1 and IL-2 from human mononuclear cells by the A chain of ML I (Franz et al. 1988). Recently, we could successfully demonstrate the release of C-reactive protein in mice.

2.4.2   Release of C-Reactive Protein

Franz et al. (1989) estimated the release of the acute phase protein CRP after peritoneal injection of ML I at sublethal and lethal doses into mice. For the quantitative estimation of CRP (i. e., of the CRP equivalent in mice), they used a technique developed by Bürger et al. (1987) explained in Fig. 2.3. This competitive enzyme ligand assay has the advantage benefit that specific antibodies against mouse-CRP are unnecessary. Figure 2.4 demonstrates a dose-dependent release of CRP by ML I. Intraperitoneal injection of 150 ng/mouse leads to a characteristic increase of CRP in serum between days 1 and 2. Later, the CRP level quickly normalizes. Interestingly, the application of 300 ng/mouse effects the CRP concentration curve with two distinct maxima at days 1 and 3 and a smaller one at day 5. Perhaps this phenomenon reflects different mechanisms of CRP release during a nonlethal intoxication by ML I. The LD_{50} for mice (intraperitoneal application) is 660 ng/20 g mouse. Injection of 1000 ng/mouse results in a rather delayed increase of the CRP level till the death of the animals at day 3. The CRP release by ML I might be due to the well known toxicity of

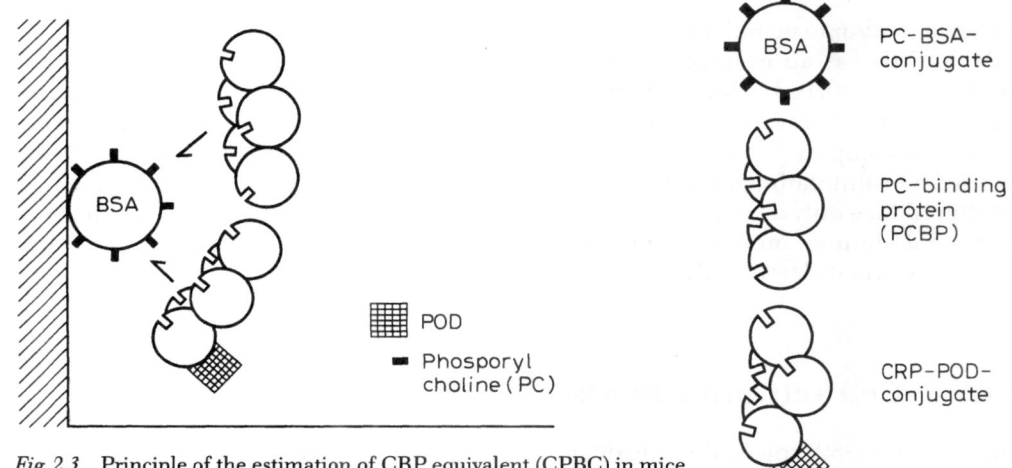

Fig. 2.3 Principle of the estimation of CRP equivalent (CPBC) in mice. CPBC and CRP (POD labeled) compete for phosphoryl choline bound to bovine serum albumin (BSA) on the solid phase

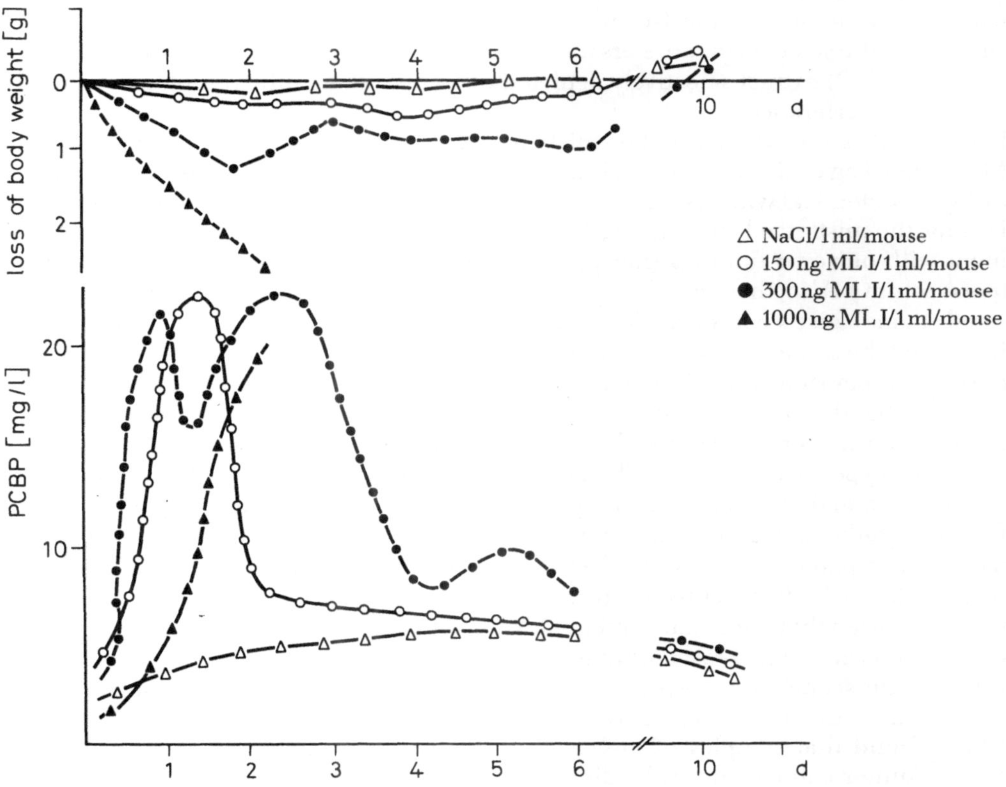

△ NaCl/1 ml/mouse
○ 150 ng ML I/1 ml/mouse
● 300 ng ML I/1 ml/mouse
▲ 1000 ng ML I/1 ml/mouse

Fig. 2.4 Estimation of mouse CPBC after i. p. injection of ML I.
After application of 150 ng a rapid increase followed by a decrease after day 2 is seen. 300 ng effect a two (or three) peak curve. After application of 1000 ng/mouse the increase continues till the death of the animal at day 2. For comparison the lost of body weight is shown

the lectin and/or to an independent mechanism. It is worth noting that the liberation of CRP by ML I is an example for the release of an endogenous lectin (CRP) by an exogenous one (ML I) with the same sugar specificity (D-galactose). In any case, these results confirm our working hypothesis that the intoxication by lectins of the ricin-type is a highly complex phenomenon. On the other hand, at low concentrations, CRP has an immunostimulating activity. Therefore, the injection of nontoxic amounts of ML I, which produce only a single phase increase of CRP, could be of therapeutical interest. The rapid elimination of CRP may be based on its fixation to galactosyl residues on phagocytes via its lectin activity.

2.4.3   Cell Activation by Viscum album Preparations

Hajto et al. (1989) injected nontoxic doses of mistletoe lectin isolated from the clinically applied extract or its B chain (0.25−1.0 ng/kg) into rabbits. They observed a significant increase in natural killer cytotoxicity, frequency of large granular lymphocytes, and phagocytic activity of granulocytes. Similar results were also obtained in cancer patients after application of Iscador (s.c. as well as i.v.). Comparative analyses of the changes in the selected parameters following injection of (1) extract with normal lectin content and (2) extract selectively depleted of lectin into healthy volunteers, corroborated this interference.

However, it is uncertain whether mistletoe preparations like Iscador really contain ML I. Holtskog et al. (1988) found in Iscador a toxic component which is closely related to, but not identical with Viscumin (ML I). Jordan and Wagner (1988) demonstrated in Iscador the mistletoe lectins II and III but not ML I. These observations are in agreement with our own findings (not yet published) and the results of Ribereau-Gayon (pers. comm.). Also, these results confirm the assumption that ML I as well as its separated chains cause a complex release of mediators (Franz 1990). It is one of our suppositions that the same complex mediator release which might have curative effects at low doses, is responsible for, or is at least involved in, the death of the animal after application of lethal doses. This would mean that the killing activity of ML I (and perhaps of ricin and abrin) is not (or not exclusively) due to the extreme cytotoxicity, but is the result of the hyperreactivity of the "body's own weapons". Hülsen et al. (1989) studied the in vitro interaction of different preparations of Helixor with human PMNC. They found a significant enhancement of natural killer cell (NK) activity in a group of tumor patients compared with a group of healthy blood donors. However they do not answer the question whether lectins are responsible for this effect. Mueller et al. (1989) recently reported that the enhancement of NK cytotoxicity by Viscum album extracts correlates well with an increased formation of lytic NK cell/tumor cell conjugates. Galacturonic acid seems to be responsible for both effects. The binding mechanism of the carbohydrate to effector cells and to target cells remains unclear. Klett and Anderer (1988) found that samples of Helixor® were unable to enhance NK cytotoxicity in PBMC/tumor cell cocultures by direct, short-term mediation but NK cytotoxicity of PMBC was strongly stimulated after preincubation with a partly purified fraction derived from exacts of Viscum album mali. The responding effector cells were identified as monocytes/macrophages. The active component of the investigated fraction has a molecular weight of about 1000 and seems to be an oligosaccharide. Because the observed activity was missing in fresh *Viscum album* extracts, the authors suggest that the active component was generated during production of the fraction and might be a result of autolytic processes induced by endogenous plant hydrolases.

Mueller and Anderer (1990) identified the component enhancing the NK cytotoxicity of human CD 56$^+$CD 3$^-$ NK cells in cocultures with K 562 tumor cells as a rhamnogalacturonan. Both activities were abolished by treatment of *Viscum album* extract with poly-α-D-galacturonidase and D-rhamnosidase and both activities were inhibited in the presence of galacturonic acid, acetylated rhamnose (6-deoxymannose), acetylated mannonic acid γ-lactone and acetylated mannose. The authors emphasize that the mode of triggering the killing mechanism by the bridging with rhamnogalacturonan must be entirely different from the mode of action of well known modifiers of spontaneous NK cytotoxicity such as IL-2 and the interferons. The reviewer does not exclude that cell surface lectins could be involved in this phenomenon and perhaps the oligosaccharide stems from the carbohydrate moiety of mistletoe lectins.

Doser et al. (1989) found two lectin fractions with almost the same cytotoxic activity on MOLT-4 cells but with different carbohydrate affinity (ML I and ML II). They inhibited the protein synthesis of MOLT-4 cells stronger than DNA synthesis.

Kohlweg et al. (1987) reported apart from a direct cytotoxicity of mistletoe extracts a "central" immunostimulation in juvenile mice (hyperplasia of the thymus, a widening of the paracortical lymphatic nodes splinomegaly).

2.4.4 Mistletoe Lectins and Breast Cancer. Histochemical Investigations

Al-Alousi et al. (1990), Al-Alousi (pers. comm.) examined the reactions between mistletoe lectins (ML I, ML II, ML III) and breast cancer cells by means of an immunoperoxidase technique. They investigated the binding of the lectins to paraffin-embedded sections of 234 patients with breast cancer followed up for 11 years. The carcinomas were classified into "stainers" (positive reaction), see Fig. 2.5, and "nonstainers". Some cases were negative for ML I, while positive for ML II and ML III. Few cases were positive for ML I while negative for ML II and ML III. All the cases (except three) that were positive for ML II were positive for ML III.

Nearly one-half of the cancer cases tested for mistletoe lectin binding showed no reactivity to any of the three lectins. No significant correlation was found between binding of these lectins and primary tumor size, blood group, tumor differentiation, tumor histology type or with lymph node status. The aim of this study was to examine the importance of these lectins as a tissue based predictive test for biological behavior in breast cancer.

With regard to the survival rate, in particular the premenopausal patients, it is worse if they were stainers for ML III than if they were stainers for ML I and ML II. The difference in survival rate between stainers and nonstainers for ML III is larger than seen with ML I and ML II.

The authors also investigated the correlation of staining with disease free interval (DFI). DFI is the time between surgery (i. e., histological confirmation of cancer) and confirmation of disease spread. Times were recorded in months. Figure 2.6 shows that 78 % of nonstainers in the postmenopausal group stayed disease-free over a period of 10 years. For stainers the rate of the disease-free state decreased from 56 % at 5 years to

40 % at 10 years. In the premenopausal group 71 % of nonstainers were disease free at 5 and 10 years. The disease-free rate for stainers was 31 % at 5 years and 18 % at 10 years. This test may be of value in predicting behavior and response to treatment in breast cancer.

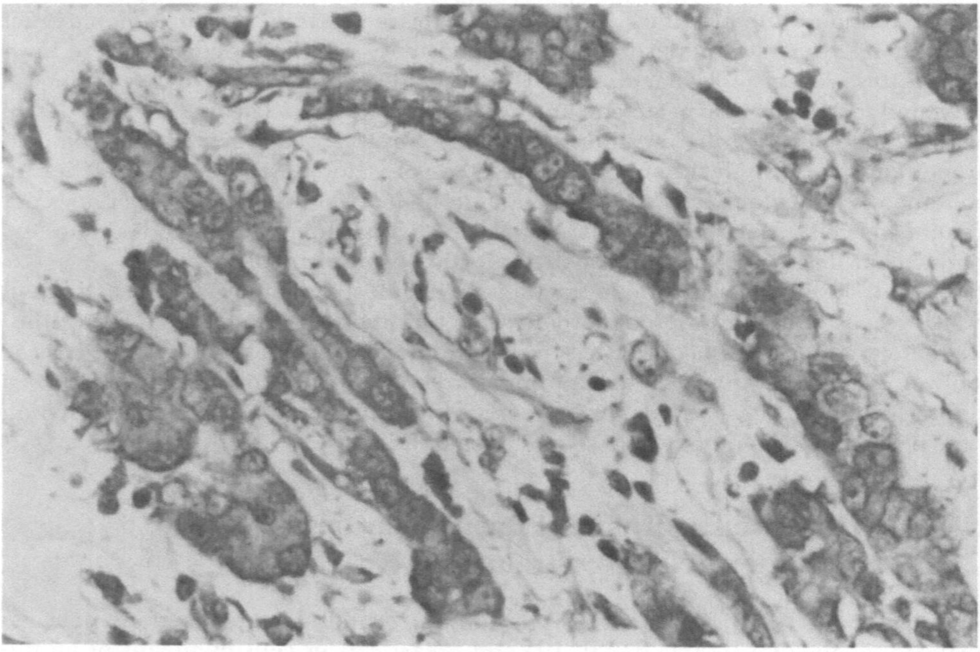

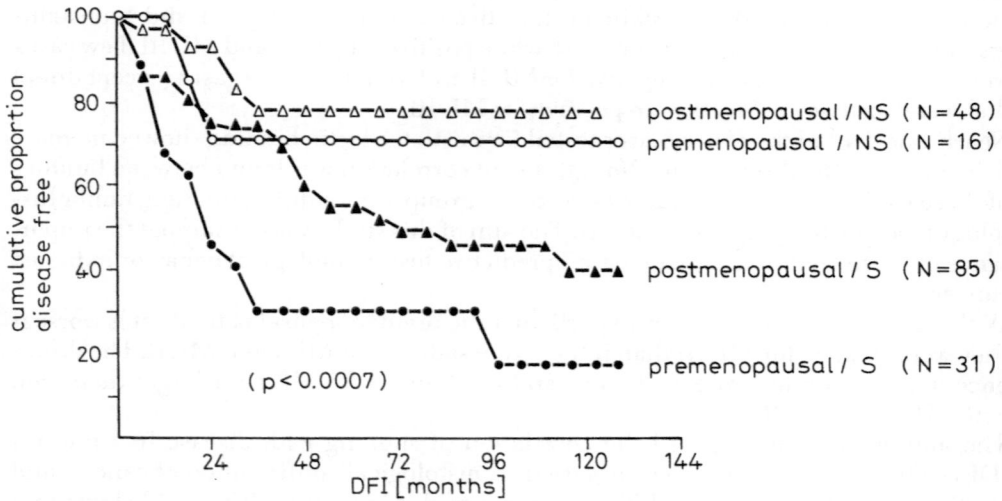

Fig. 2.5 Human primary breast cancer. Tumor cells and lymphocytes are positive for ML I

Fig. 2.6 Disease free interval (DFI) of different groups of patients with primary breast carcinoma (▲● stainers. △○ non stainers for ML III). For explanation see paragraph 2.4.4

2.4.5 Immunotoxins Containing the A Chain of ML I

The first immunotoxin-like conjugate consisting of histamine and the A chain of ML I was described some years ago by Eckert et al. (1985). This "affinotoxin" attacks cells with histamine receptors on their surface.

Schütt et al. (1989) prepared an immunotoxin consisting of a human monocyte specific monoclonal antibody (RoMo-1) and partially denatured ML I (dML I). For this purpose ML I was exposed to salts with melting points below 20 °C (so-called molten or fused salt liquid at room temperature). After treatment by ethylammonium nitrate (melting point 12 °C) the B chain lost its sugar-binding activity, but the A chain remained ribosome-inactivating. It could be observed that in vitro exposure for 24 h to this immunotoxin was associated with the disappearance of monocytes from mononuclear cells (MNC) evaluated as RoMo-1-positive cells, whereas no selective removal of monocytes was observed in cultures exposed to RoMo-1 and dML I. Moreover, the PHA reactivity of T-cells in a MNC suspension is significantly diminished after immunotoxin interaction, but can be reconstituted by addition of MNC in relatively small amounts (Table 2.2). The use of toxic lectins partially denatured by means of molten salts might be a new variant for preparing immunotoxins.

Table 2.2 *In vitro* selectivity for monocytes of RoMo-1-dML I immunotoxin (IT) as compared to dML I and RoMo-1

Experiment	μl/ml	% Monocytes* 0 h	24 h	0.5 h	% Dead cells[†] 4 h	7 h	24 h	PHA reactivity (counts/min $\times 10^{-3}$ ^{3}H-TdR)
Control medium	–	19	18	2.4	2.7	2.2	1.9	10.3 ± 2.4
Control RoMo-1	15.0	20	17	2.1	3.0	2.2	4.7	9.7 ± 3.0
IT	27.0	19	0	1.1	4.4	0.1	10.7	0.3 ± 0.1
dMLI	14.0	19	17	1.8	2.0	2.0	2.0	12.0 ± 1.8

Blood MNC (5×10^6/ml) were treated with IT, dMLI and RoMo-1 for 24 h, washed and then cultured for 72 h in the presence of PHA. Monocytes proportion is expressed in terms of FITC-RoMo-1-positive cells (*S.D.<15 %, triplicates), dead cells proportion is expressed as ethidiumbromide-positive cells ([†]S.D.<25 %, triplicates). Mol. wt. of IT is about 300,000.

2.5 Lectin/Sugar Interactions in Nonpolar Solvents

Up to now lectin-sugar (glycoconjugate) interactions have been performed in aqueous systems. Pfüller et al. (1989) investigated the reaction of ML I and other lectins (ricin, *Lens culinaris* lectin, Con A) with carbohydrates in iso-octane. They entrapped the lectins in reverse micelles formed by the anionic surfactant Aerosol OT (AOT) in iso-octane after addition of small amounts of water. The core of the reverse micelles consists of the polar head groups of AOT, the counter ions, and a water pool containing the lectins. For testing the sugar binding activity of the entrapped lectins in iso-octane, the authors chose the corresponding sugars covalently bound to controlled pore glass particles (CPG/100 A). In the presence of inhibiting sugars, such as 4-nitrophenyl-α-D-galactosides and α-glucosides in the water pool, the binding of the lectins to the carrier was inhibited.

The amount of the sugar bound lectins was estimated quantitatively according to Bradford (Table 2.3, column A). Also, the measurement of the increased fluorescence of dansylated amino sugars after binding to lectins indicates an interaction of these sugars with lectins in the apolar system, comparable to that in an aqueous system. The different behavior of Con A needs further studies. Perhaps these results will lead to the application of self-organized surfactant assemblies (e. g., liposomes, microemulsions, liquid crystals) in lectin research.

Table 2.3 Comparison of lectin-sugar interactions in aqueous and apolar systems

Lectin	A[a]	Dansylated sugar	B[b] Fluorescence increase %
ML I	95%	Dans-Gal N	80
Ricin	84%	Dans-Gal N	83
Con A	2%	Dans-Glc N	10
LCA	70%	Dans-Glc N	70

[a] Column A: quotient of CPG-sugar-bound lectin in AOT/iso-octane to saline.
[b] Column B: ratio of fluorescence increase of dansylated amino sugars after binding to lectins, binding in AOT/iso-octane and saline, respectively. For further explanation see paragraph 2.4.6

Table 2.4 Quantification of ligand release from ML I-Sepharose 4B by solid phase EIA calculated as ML I liberated (ng/ml) per mg ML I coupled

No. of elution buffer	Elution buffer	Sugar in the elution buffer (mol/l)	ML I-release (ng ML I/mg · ml) ± sd
1	PBS, pH 7.4	0	14.5 ± 0.5
2	PBS, pH 7.4	0.01 D-gluc	12.1 ± 0.7
3	PBS, pH 7.4	0.05 D-gluc	23.4 ± 1.7
4	PBS, pH 7.4	0.1 D-gluc	26.8 ± 1.2
5	PBS, pH 7.4	0.5 D-gluc	54.1 ± 2.4
6	PBS, pH 7.4	0.5 M NaCl	23.5 ± 1.6
7	PBS, pH 7.4	0.01 D-gal	20.2 ± 0.9
8	PBS, pH 7.4	0.05 D-gal	21.6 ± 1.1
9	PBS, pH 7.4	0.1 D-gal	24.5 ± 1.0
10	PBS, pH 7.4	0.5 D-gal	60.3 ± 2.6
11	glycine/HCl, pH 2.5	0	17.8 ± 0.8

2.6 Ligand Leakage from ML I Affinity Columns

Walzel et al. (1989) investigated the leakage of ML I from columns prepared for affinity chromatography. Traces of ML I in glycoproteins isolated by ML I-Sepharose 4 B affinity chromatography have been demonstrated by immunoblotting. The quantitative estimation of ML I release was performed by solid phase EIA (Table 2.4). Ligand release was also observed from glutaraldehyde stabilized affinity gel. This could be due to a low content of lysine residues in ML I, which results in inefficient cross-linking. The relatively high degree of leakage might be explained by the binding of dimers to the beads with binding of only one molecule to the solid phase. By means of ML I columns for purification of glycoconjugates a further fine purification of glycoconjugates by immunoaffinity chromatography (anti ML I-Sepharose) is recommended.

2.7 Conformation of ML I at Different pH

The conformation of ML I at different pH has been investigated by Bushueva et al. (1988) using intrinsic fluorescence measurement. The treatment of ML I with a denaturant (guanidine hydrochloride) and with quenchers of the intrinsic fluorescence (I^-, Cs^+, acrylamide) indicates different structures at pH 7 and 4. At pH 4 tryptophan residues become more accessible to quenchers. The positive charge of the surrounding area increases and the lectin becomes more stable to the action of guanidine hydrochloride. The stability of the isolated chains of ML I differs considerably from that of the complete lectin. Both A- and B-chains are more sensitive against guanidine hydrochloride. The stability of the isolated chains depends on the ionic strength of the solvent. Moreover, the B-chain displays an increased accessibility of tryptophan residues to quenchers. Differences between the conformations of the isolated chains at pH 7 and 4 are marked more strongly. At pH 4.5 the B-chain undergoes a structural transition which is possibly related to its auxiliary function in the membrane transfer of ML I. Similar results have been described for ricin by Bushueva and Tonevitsky (1987).

2.8 Further Characterization of ML I, ML II, and ML III

The occurrence of three different lectins in dried mistletoe plants was described by Franz et al. in 1981. The three lectins differ in their molecular weight and in sugar specificity. In a recent paper Ziska et al. (1989) compared in detail the sugar specificity of ML I with that of ML II and ML III, respectively.

ML II and ML III have been purified using FPLC ion exchange chromatography on a Mono S cation exchanger. The sugar specificity was estimated using the hemagglutination inhibition test (Table 2.5). Furthermore ML I, ML II and ML III are not inhibited by D-glucose, D-glucosamine, N-acetyl-D-glucosamine, D-mannose, D-mannosamine, N-acetyl-D-mannosamine, maltose, or sucrose (0.02 M).

Among the inhibiting D-galactosides, differing only in the configuration at the anomeric C-1 atom, there seems to be a slight preference for β-anomers. Aromatic aglycons of D-galactosides increased binding of the corresponding sugars to all lectins. Inhibition studies with ML I and D-galactose derivatives demonstrated that substitution of any of the hydroxyl groups at positions C-2, C-3 and C-4 of the D-galac-

Table 2.5 Inhibition power of D-galactose derivatives: μmoles/ml needed for complete inhibition of a lectin solution with a titer of 1:8

Inhibiting compound	ML-I	ML-II	ML-III
D-Galactose	12	12	50
N-Acetyl-D-galactosamine	200	6	3
Methyl-α-D-galactopyranoside	12	6	12
Methyl-β-D-galactopyranoside	6	3	6
Phenyl-α-D-galactopyranoside	3	6	6
Phenyl-β-D-galactopyranoside	1.5	3	3
2-Deoxy-D-galactose	200	25	25
2-O-Methyl-D-galactose	200	25	25
3-O-Methyl-D-galactose	200	200	200
6-O-Methyl-D-galactose	12	25	25
6-Deoxy-D-methyl-galactose	12	25	25
Lactose (β-1-4)	3	3	3
Lactulose (β-1-4)	3	3	3
Melibiose (α-1-6)	12	12	12
Raffinose (α-1-6, β-1-2)	12	12	12
D-Talose (C-2 epimer)	200	200	200
D-Gulose (C-3 epimer)	200	200	200

topyranosyl ring system abolished the binding to the lectin. Hemagglutination inhibition tests demonstrated very similar lectin-carbohydrate binding activities. The fact that 3-O-methyl-D-galactose, the C-3 epimeric D-gulose and the C-4 epimeric D-glucose did not inhibit the hemagglutination by ML II and ML III points to a requirement for a free equatorial hydroxyl group at C-3 and an axial one in position C-4; the C-6 hydroxyl group does not seem to be essential for binding D-galactose since 6-deoxy-D-galactose and 6-O-methyl-D-galactose do not act as inhibitors.

In contrast to ML I, ML II and ML III do not need any free hydroxyl group in position C-2. This is supported by the fact that 2-deoxy-D-galactose and N-acetyl-D-galactosamine inhibited hemagglutination. None of the three lectins, however, showed a reaction with C-2 epimeric D-talose. From inhibition data it is assumed that the hydroxyl configuration at the C-2, C-3, and C-4 atoms is structurally the most critical position for successful saccharide binding.

The carbohydrate binding specificity for the lectins of mistletoe is summarized as follows: ML I D-galactose ≫ N-acetyl-D-galactosamine; ML II D-galactose ~ N-acetyl-D-galactosamine and ML III D-galactose < N-acetyl-D-galactosamine.

The authors also determined the toxicity of the lectins after intraperitoneal injection into mice (LD_{50}). The most toxic one is ML I (LD_{50} 28μg/kg), followed by ML III (55μg/kg) and ML II (1450μg/kg). It must be pointed out that ML III is the first toxic lectin of the AB type containing a B chain with a dominant affinity to N-acetyl-galactosamine. Ricin, abrin, mistletoe lectin I, modeccin, and the lectin from *Phoradendron californicum* contain B chains with D-galactose specificity. Doser (1988) investigated the lectin content in commercial *Viscum album* preparations (ABNOBA, Helixor). He found mainly ML 1 (ML I) in ABNOBA[R] and ML 2 (presumably indentical to ML III as described by Franz et al. [1981]) in Helixor. The lectin estimation in aqueous extracts from mistletoes grown on different host trees resulted only in small differences. Doser did not exclude an alteration of the lectins during the extraction. First, the cytotoxicity can be diminished. He found that the total cytotoxicity of plant extracts correlated with cytotoxicity. Finally, Doser concluded that from a therapeutical point of view the cytotoxicity is the most important biological activity of *Viscum album* extracts.

In regard to the existence of different mistletoe lectins with different sugar binding activities, I would like to formulate a hypothesis: The B-chain of mistletoe lectin(s) contains two different binding sites, one for D-Gal and one for D-GalNac. Alterations of the conformation of the B chain perhaps caused by splitting off the peptides (or oligosaccharides?) favor either the Gal- or the Gal-Nac-binding site. This hypothesis could explain the existence of *Viscum album* lectins with different affinities to Gal/GalNac and also differences in cytotoxicity of mistletoe extracts in depending on the production procedure.

2.9 Concluding Remarks

The recent development in mistletoe lectin research is characterized by an increasing interest in the biological activities of the lectins, their chains, and also in their fragments. In the next years the elucidation of the primary structure of ML I might be expected. Moreover, cytotoxicity and complex mediator release are worthy of note. According to our present knowledge, active centers of the ML I molecule are the enzymatic site of the A chain, the sugar binding locus in the B chain, hydrophobic sites on both chains, and perhaps also their sugar moieties. The diversity of their biological activities renders lectins from *Viscum album* to be potential drugs.

2.10 References

Al-Alousi M, Leatham A, Young T, Franz H (1990) Mistletoe lectins and behaviour of breast cancer. Proc Roy microsc Soc 25:52

Bürger W, Ritter E, von Baehr R (1987) Ein schneller quantitativer Test für C-reaktives Protein im Serum. Z Klin Med 42:1999–2001

Bushueva TL, Tonevitsky AG (1987) The effect of pH on the conformation and stability of the structure of plant toxin-ricin. FEBS-Lett 215:155–159

Bushueva TL, Tonevitsky AG, Kindt A, Franz H (1988) Структура токсичного белка-лектина из омелы при различных pH: исследование методом собственной флюоресценции. Mol Biol (Moscow) 22:628–634

Bushueva TL, Tonevitsky AG, Kindt A, Franz H (1990) The structure of mistletoe lectin I at different pH: Studies by the method of intrinsic fluorescence. In: Kocourek J (ed) Proc 10th Lectin Meet, Prague, ČSFR 1988 Lectins: biology, biochemistry, clinical biochemistry, vol 7. Sigma Chemical Company St. Louis, Missouri, USA pp 179–185

Doser M (1988) Ermittlung zellbiologischer Wirkungen verschiedener Fraktionen aus Extrakten von Viscum album L. Thesis, Hohenheim

Doser C, Doser M, Hülsen H, Mechelke F (1989) Influence of carbohydrates on the cytotoxicity of an aqueous mistletoe drug and of purified mistletoe lectins tested on human T-leukemia cells. Arzneim-Forsch/Drug Res 39 (I): 647–651

Eckert R, Pfüller U, Kindt A, Reichelt E, Franz H (1985) Histaminrezeptortragende Lymphozyten III. Suppression von Immunreaktionen nach Abtöten Histaminrezeptor-tragender Lymphozyten durch ein Konjugat aus Histamin und der A-Kette des Mistellektins I. Biomed biochim acta 44:1239–1245

Franz H (1989a) Viscaceae lectins. In: Franz H (ed) Advances in lectin research, vol 2. Volk und Gesundheit, Berlin, pp 28–59

Franz H (1989b) The in vivo toxicity of toxic lectins is a complex phenomenon. In: Kallikorm A, Bøg-Hansen TC (eds) Proc 11th Int Lectin Meet, Tallinn, Estonia 1989 Lectins: biology, biochemistry, clinical biochemistry, vol 8. Sigma Chemical Company St. Louis, Missouri, USA, in press

Franz H (1990a) 100 Jahre Lektinforschung – eine Bilanz. Naturwissenschaften 77:103–109

Franz H (1990b) The use of mistletoe lectin I in brain histochemistry. Ergebn exp Med 51:216–224

Franz H, Ziska P, Kindt A (1981) Isolation and properties of three lectins from mistletoe (Viscum album L.). Biochem J 195:481–484

Franz H, Friemel H, Buchwald S, Plantikow A, Kopp J, Körner I-J, (1990) The A chain of lectin I from European mistletoe (Viscum album) induces interleukin-1 and interleukin-2 in human mononuclear cells. In: Kocourek J (ed) Proc 10th Int Lectin Meet, Prague, ČSFR 1988 Lectins: biology, biochemistry, clinical biochemistry, vol 7. Sigma Chemical Company St. Louis, Missouri, USA pp 247–250

Franz H, Pfüller K, Bürger W (1989) Mistletoe lectin I (ML I) releases C-reactive protein in mice. In: Kallikorm A, Bøg-Hansen TC (eds) Proc 11th Int Lectin Meet, Tallinn, Estonia 1989 Lectins: biology, biochemistry, clinical biochemistry, vol 8. Sigma Chemical Company St. Louis, Missouri, USA, in press

Gossrau R, Franz H (1990) Histochemical response of mice to mistletoe lectin I (ML I). Histochemistry 94:531–537

Gossrau R, Graf R, Günther T, Merker H-J, Nau H, Stahlmann R (1988a) Enzyme cytochemistry for the risk assessment of drug administration during pregnancy with special reference to proteases. Acta histochem Suppl 36:361–376

Gossrau R, Günther T, Merker H-J, Graf R (1988b) Enhancement of maternal and fetal nephrotoxicity of salicylate by zinc deficiency. Morphological, enzyme histochemical, and immunohistochemical studies. Histochemistry 89:81–90

Hajto T, Hostanska K, Gabius H-J (1989) Modulatory potency of the β-galactosidespecific lectin from mistletoe extract (Iscador) on the host defense system in vivo in rabbits and patients. Cancer Res 49:4803–4808

Hajto T, Hostanska K, Gabius H-J (1990) Zytokine als Lektin-induzierte Mediatoren in der Misteltherapie. therapeutikon 4:136–145

Holtskog R, Sandvig K, Olsnes S (1988) Characterization of a toxic lectin in Iscador, a mistletoe preparation with alleged cancerostatic properties. Oncology 45:172–179

Hülsen H, Kron R, Mechelke F (1989) Influence of Viscum album preparations on the natural killer cell-mediated cytotoxicity of peripheral blood. Naturwissenschaften 76:530–531

Jordan E, Wagner H (1986) Nachweis und quantitative Bestimmung von Lektinen und Viscotoxinen in Mistelpräparaten. Arnzeim-Forsch/Drug Res 36 (I): 428–433

Klett CY, Anderer FA (1989) Activation of natural killer cell cytotoxicity of human blood monocytes by a low molecular weight component from Viscum album extracts. Arzneim-Forsch/Drug Res 39 (II): 1580–1585

Kohlweg EJ, Dijno A, Gallent E, Schussnig R, Hammerschlag R (1987) In vitro-Nachweis der lymphozytenstimulierenden Eigenschaft von Extrakten aus Viscum album mali. Krebsgeschehen 19: 1–6

Metzner-G, Franz H, Kindt A, Fahlbusch B, Süss J (1985) The in vitro activity of lectin I from mistletoe (ML I) and its isolated A- and B-chains on functions of macrophages and polymorphonuclear cells. Immunobiology 169:461–471

Mueller EA, Anderer FA (1990) Chemical specificity of effector cell/tumor cell bridging by a Viscum album rhamnogalacturonan enhancing cytotoxicity of human NK cells. Immunopharmacology 19:69–71

Mueller EA, Hamprecht K, Anderer FA (1989) Biochemical characterization of a component in extracts of Viscum album enhancing human NK cytotoxicity. Immunopharmacology 17:11–18

Pfüller U, Franz H, Müller P (1989) Lectin-sugar interaction in selforganized surfactant assemblies. In: Kallikorm A, Bøg-Hansen TC (eds) Proc 11th Int Lectin Meet, Tallinn, Estonia 1989 Lectins: biology, biochemistry, clinical biochemistry, vol 8. Sigma Chemical Company St. Louis, Missouri, USA, in press

Ritchie DG (1990) Interleukin 6 stimulates hepatic glucose release from prelabeled glycogen pools. Am J Physiol 258 (Endocrinol Metab 21): E 57–E 64

Schütt C, Pfüller U, Siegl E, Walzel H, Franz H (1989) Selective killing of human monocytes by an immunotoxin containing partially denatured mistletoe lectin I. Int J Immunopharmac 11:977–980

Walzel H, Mix E, Jenssen H-L, Ziska P, Franz H (1982) Estimation of toxicity of the mistletoe lectin I using the foodpad swelling test in mice. Acta histochem 71:41–42

Walzel H, Brock J, Franz H, Ziska P (1989) Detection and quantification of ligand leakage from lectin affinity columns. Biomed biochim acta 48:221–226

Yoshida T, Zhang M, Chen C, Franz H, Wu HC (1991) Enhancement of the cytotoxicity of mistletoe lectin I (ML I) by high pH or pertubation in Golgi functions. Pharmazie, in press

Ziska P, Gelbin M, Franz H (1989) Interaction of mistletoe lectins ML I, ML II, and ML III with carbohydrates. In: Kallikorm A, Bøg-Hansen TC (eds) Proc 11th Int Lectin Meet, Tallinn, Estonia 1989 Lectins: biology, biochemistry, clinical biochemistry, vol 8. Sigma Chemical Company St. Louis, Missouri, USA, in press

Additions in Print

Stettin et al. (1980) investigated the humoral response to an aqueous mistletoe extract in 23 tumor patients after treatment from 2 month up to 6 years with increasing dosages. Antibodies of IgG class have been demonstrated by several immunological techniques using both the mistletoe preparation and purified ML I as antigenic components. No IgM antibodies have been found. Immunoblotting revealed lectin specific antigens at 62,000 (ML I) 33,000 (B chain) and 29,000 (A chain) MW. The inhibition of PHA induced proliferation of normal lymphocytes was decreased by anti ML I positive sera. Perhaps antibodies against mistletoe lectins protect patients from diserable cytotoxic effects of the lectins. On the other hand it is remarkable that till now no side reactions caused by antigen-antibody complexes are described. That might be the follow of the complex formation of mistletoe lectins with glycoconjugates of the serum.

Jung et al. (1990) compared cytotoxic proteins from mistletoe extracts with those of the therapeutically used preparation Iscador[R]. By cation exchange chromatography cytotoxic proteins from the mistletoe extract were mainly eluted at the same positions as purified lectins while those of Iscador[R] were eluted at the position of viscotoxins. The authors point out that ML I, the main cytotoxic protein of *Viscum album* represents only a small part of the total cytotoxic proteins in Iscador[R].

Sauviat (1990) detected that Phoratoxin B, a toxin isolated from mistletoe has a depolarizing action on frog skeletal muscle fibres. This effect is based on increasing the resting membrane conductance. It seems to exist an interaction between Ca^{++} and the toxin molecule at the membrane level. Perhaps the toxic one acts as a detergent, similar to cardiotoxins.

In order to investigate the influence of serum glycoconjugates on the binding of ML I to cell membrane receptors, Sörgel (1991) incubated cryostate sections of mamma carcinoma with FITC-labelled ML I and with a mixture of the labelled lectin and human serum. He found a significant decrease of the fluorescence intensity but not a complete loss of fluorescence. These results indicate a competition of cell receptors and serum glycoconjugates for ML I. On the other hand, the results underline the relatively high affinity of mamma carcinoma receptors to ML I. Our earlier results revealed a complete inhibition of erythrocyte agglutination by human serum.

Kießig et al. (1990, in prep.) treated mononuclear cells from peripheral blood of healthy donors (1×10^6 cells, 200 µl/well in RPMI with 10 % fetal calf serum on microtest plates 200 µl per well at 37 °C) with decreasing amounts of ML I and its A chain, respectively. Both the pokeweed mitogen

(PWM) stimulated and the spontaneous synthesis (without PWM) were tested. IgM in the supernatant was estimated by ELISA.

Hajto et al. (1990) described an increased secretion of tumor necrosis factor α, Interleukin 1 and Interleukin 2 from human mononuclear cells by ML I. The same effect they found for the galactoside-binding subunit (B chain). These activities might be responsible for immunomodulatory potency of ML I. This is partially in agreement with the results of Männel et al. (1991) and with results obtained together with G. Metzner (Franz 1990). Northern blot investigations demonstrated that ML I as well as its A and B chain release from human mononuclear cells beside IL 1, TNF and IL 6 also IL 8 and MCAF (macrophage chemotoxis activating factor). The corresponding activity of the A chain which is not a galactose specific lectin but a N-glycosidase indicates that the mediator-releasing effect of ML I is not necessarily due to its galactose binding activity.

Ziska et al. (1985) described a procedure for the preparation of mistletoe extracts with a defined content of lectins for use in therapy, mainly in cancer treatment. Using ultrafiltration with different membranes a lectin-free fraction was received. By addition of concentrated fresh extracts of *Viscum album* with a determined lectin content the lectin concentration of the final solution was adjusted. To estimate the lectin concentration an ELISA technique with monoclonal antibodies was used. The method has the benefit, that also the concentration of low molecular weight substances (e. g. viscotoxins) can be regulated.

In agreement with the demonstration of microglia in the brain of rodents (see above) Niedobietek and Franz (1990) demonstrated macrophages by means of ML I in the brain of patients with HIV encephalitis mainly in the perivascular space (see Figure).

Debray and Franz (1990) analyzed the carbohydrate chains isolated from ML I. Glycopeptides

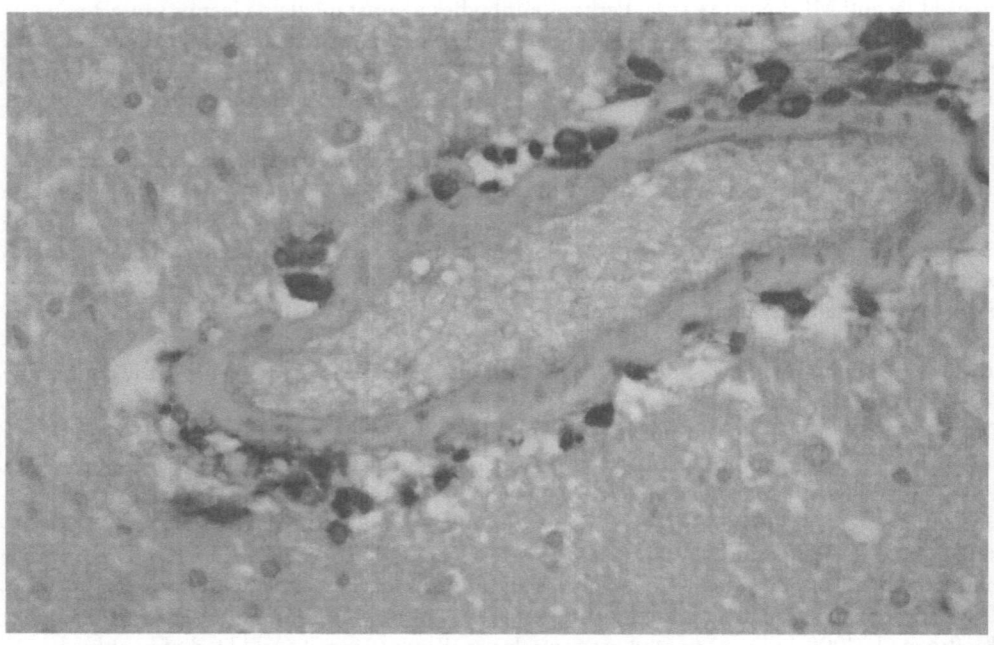

Demonstration of macrophages in the brain of patients with HIV encephalitis in the perivascular space by means of ML I/ABC technique

prepared from mistletoe lectin I by pronase digestion, were fractionated by affinity chromatography on a Concanavalin A-Sepharose column. A weakly-reactive, retarded fraction (Fraction I) and a strongly-reactive fraction eluted with an 0.5 M solution of methylα-D-glucoside (Fraction II), were obtained. The carbohydrate composition of the two glycopeptide fractions suggests that D-xylose-containing glycopeptides from Fraction I possess glycan structures similar to those found in other plant glycoproteins. On the contrary, Con A strongly-reactive glycopeptides from Fraction II, which contain only mannose and N-acetylglucosamine residues, seem to possess oligomannosidic-type glycans.

^{1}H-chemical shifts of structural-reporter groups of the constituent monosaccharides for the two glycopeptide fractions were established by 400-MHz ^{1}H-NMR spectroscopy.

The ^{1}H-NMR data of the D-xylose-containing glycopeptide fraction (Fraction I) demonstrated one N-linked heptasaccharide structure, which corresponds to the common pentasaccharidic inner-core of N-glycosylproteins, with the occurrence of a β-1,2-linked xylose residue to the β-mannosyl residue of the mannotriose core and a fucose residue α-1,3-linked to the N-acetylglycosamine residue involved in the N-glycosylamine linkage (Structure 1). The same structure has been demonstrated to occur in lectins from *Clerodendron trichotomum*, from five *Erythrina* species, from *Sophora japonica*, from *Lonchocarpus capassa*, from *Ricinus communis* and from *Artocarpus integrifolia*.

The set of chemical shift values of the structural reporter groups obtained for the Con A strongly-reactive glycopeptide Fraction II is consistent with an oligomannosidic-type glycan containing six mannose and two N-acetylglucosamine residues (Structure 2). The same glycan has been demonstrated to occur in ricin.

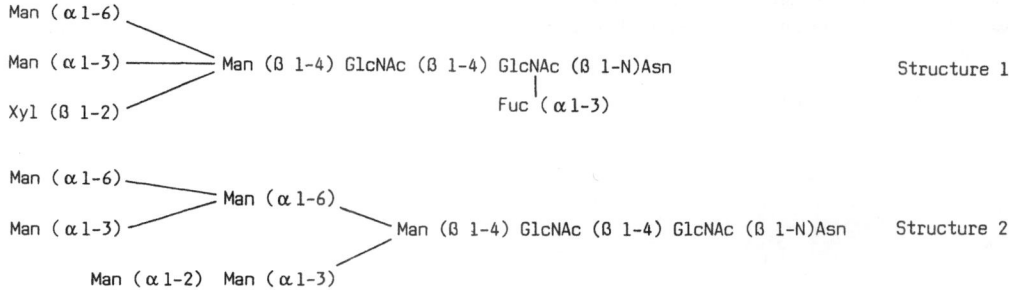

The same authors studied the affinity of ML I toward various N-linked glycopeptides and related compounds.

The behavior of N-acetyllactosamine-type oligosaccharides and glycopeptides on a column of mistletoe lectin I immobilized on Sepharose was examined. The immobilized lectin does not show any affinity for asialo-N-glycosylpeptides and related oligosaccharides, which possess one to four unmasked N-acetyllactosamine sequences. However, substitution of at least one of the N-acetyllactosamine sequences by sialic acid residue, either at C-3 or C-6 of galactose slightly enhances the affinity of the lectin. Such sialylated N-glycosylpeptides or oligosaccharides are eluted from the lectin column by the starting buffer as retarded fractions.

Surprisingly, the affinity of the immobilized mistletoe lectin I is higher for pentaantennary N-glycosylpeptides with P_1-serologic activity, isolated from turtle-dove ovomucoid. These glycopeptides possess two unmasked N-acetyllactosamine sequences and two N-acetyllactosamine sequences substituted each by an α-1,4-linked galactose residue and one such sequence substituted by an α-NeuAc residue. Such glycans are strongly bound on the lectin column and their elution is obtained with 0.15 M-galactose in the starting buffer.

References

Debray H, Franz H (1990) Manuscripts in preparation.

Hajto T, Hostanska K, Frei K, Rordorf C, Gabius HJ (1990) Increased secretion of tumor necrosis factor, interleukin 1, and interleukin 6 by human mononuclear cells exposed to -galactoside-specific lectin from clinically applied mistletoe extract. Cancer Res 50:3322−3326

Jung ML, Baudino S, Ribéreau-Gayon C, Beck JP (1990) Characterization of cytotoxic proteins from mistletoe (*Viscum album* L.). Cancer Lett 51:103−108

Kießig ST, Adrian K, Franz H (1990) Effekte von Mistellektin I auf die IgM-Synthese peripherer Blutlymphozyten. (in prep.)

Männel D N, Becker H, Gundt A, Kist A, Franz H (1991) Induction of tumor necrosis factor expression by a lectin (ML I) from *Viscum album*. Cancer Immunol Immunother, in press

Niedobietek F, Franz H (1990) Manuscript in preparation.

Sauviat MP (1990) Effect of phoratoxin B, a toxin isolated from mistletoe, on frog skeletal muscle fibres. Toxicon 28:83−89

Sörgel F (1991) Thesis, Univ Leipzig

Stettin A, Schultze JL, Stechemesser E, Berg PA (1990) Anti-mistletoe lectin antibodies are produced in patients during therapy with an aqueous mistletoe extract derived from *Viscum album* L. and neutralize lectin-induced cytotoxicity in vitro. Klin Wochenschr 68:896−900

Ziska P, Franz H, Pfüller K, Zschoche S (1985) Verfahren zur Gewinnung von krebshemmenden Mistelextrakten mit definiertem Lektingehalt. Patentschrift DD 235 418 A1, Deutsche Demokratische Republik, 19.03.85

3 Lectin Affinity Chromatography of Glycoconjugates

H. Debray and J. Montreuil

3.1 Introduction

3.1.1 Type of Glycans That Lectins Can Recognize at the Cell Surface

Glycoproteins result from the covalent association of carbohydrate moieties (glycans) with proteins. Glycans are conjugated to peptide chains through two main types of covalent linkages (N-glycosyl and O-glycosyl) leading to two classes of glycoproteins: N-glycosylproteins and O-glycosylproteins (for recent reviews, see Montreuil 1980, 1982, 1984 a, b; Kobata 1984; Kornfeld and Kornfeld 1985).

The N-glycosidic bond found in N-glycosylproteins is N-acetylglucosaminyl-asparagine/:GlcNAc β-1, N-Asn. In contrast, the O-glycosidic bond displays a wide variety of linkages, the most common being the mucin type: an alkali-labile linkage between N-acetylgalactosamine and L-serine or L-threonine: GalNAc α-1,3-Ser/Thr, found in numerous glycoproteins (mucin-type glycoproteins) including plasma cell membranes and glycoproteins from biological fluids. Some examples of structures of glycans O-glycosidically conjugated to protein through the GalNAcα-1,3-Ser/Thr linkage are shown in Fig. 3.1.

The N-glycosylproteins are divided into three families according to the nature of the carbohydrate moiety linked to their common pentasaccharidic inner core. In the first family, the glycans contain mannose and N-acetylglucosamine only. The are called oligomannosidic-type glycans or high mannose type glycans (Fig. 3.2). In the second family, glycans contain galactose, fucose and sialic acid in addition to mannose and N-acetylglucosamine. They fundamentally derive from the addition to the pentasaccharidic inner core of a variable number of N-acetyllactosamine residues/:Gal β-1,4-GlcNAc. These structures are called N-acetyllactosamine type or complex type glycans (Fig. 3.3). Glycans containing linear or branched repeating N-acetyllactosaminyl sequences have been found and are called poly(glycosyl)peptides or poly(N-acetyllactosamine)glycans (Fig. 3.4).

In the third family, glycans contain both structures of the oligomannosidic and N-acetyllactosamine types and are called hybrid type glycans (Fig. 3.5). This brief survey shows that even if both N- and O-glycosidically linked glycans possess a common inner core, they display an extreme heterogeneity due to the number and positions of the most externally situated monosaccharides in the glycan. This heterogeneity still represents one of the most difficult problems encountered in the primary structure determination of glycans. However, immobilized lectins can now be successfully used for fractionation and for structural studies mainly of asparagine-linked glycans.

51

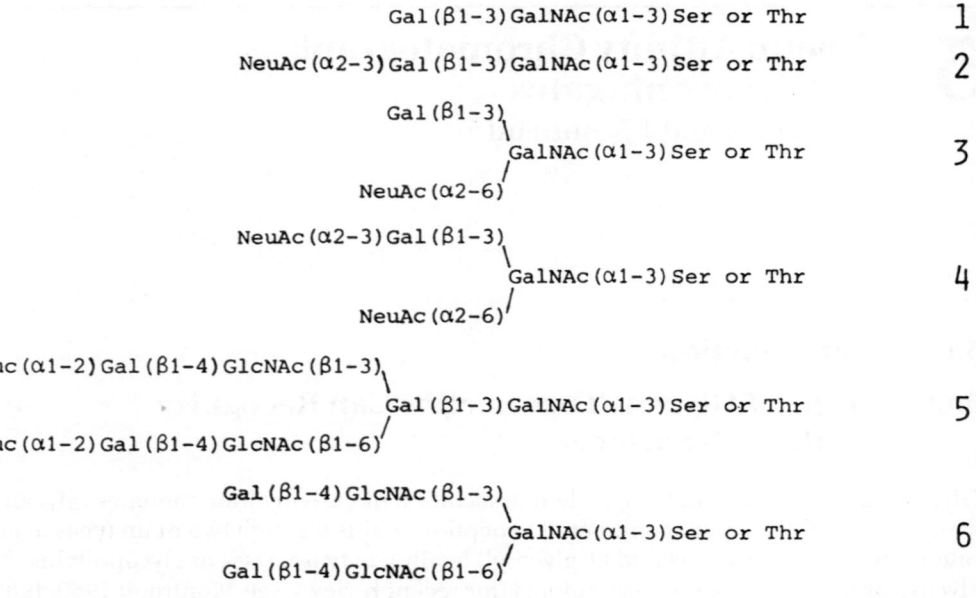

Gal (β1–3) GalNAc (α1–3) Ser or Thr   **1**

NeuAc (α2–3) Gal (β1–3) GalNAc (α1–3) Ser or Thr   **2**

Gal (β1–3)⟍
              GalNAc (α1–3) Ser or Thr   **3**
NeuAc (α2–6)⟋

NeuAc (α2–3) Gal (β1–3)⟍
                   GalNAc (α1–3) Ser or Thr   **4**
NeuAc (α2–6)⟋

Fuc (α1–2) Gal (β1–4) GlcNAc (β1–3)⟍
                          Gal (β1–3) GalNAc (α1–3) Ser or Thr   **5**
Fuc (α1–2) Gal (β1–4) GlcNAc (β1–6)⟋

Gal (β1–4) GlcNAc (β1–3)⟍
                 GalNAc (α1–3) Ser or Thr   **6**
Gal (β1–4) GlcNAc (β1–6)⟋

Fig. 3.1 Structure of glycans O-glycosidically conjugated to protein through the GalNAc α-1.3-Ser/Thr linkage.
Structure 1: anti freeze glycoprotein from Antarctic fish, human chorionic gonadotropin β-subunit, human antifreeze serum IgA₁, epiglycanin of TA 3-Ha cells, T-reactive erythrocytes, rat brain glycoproteins. Structure 2: human glycophorin, bovine Kappa-casein, fetuin, human milk secretory IgA, N-blood group. Structure 3: submaxillary mucins, rat brain glycoproteins, trout and herring eggs. Structure 4: human glycophorin and gonadotropin β-subunit, fetuin, lymphocyte plasma membrane, M-blood group. Structure 5: porcine blood group H substance. Structure 6: bronchial mucin of patients suffering from cystic fibrosis (for reviews, see Montreuil 1980, 1982, 1984 a, b)

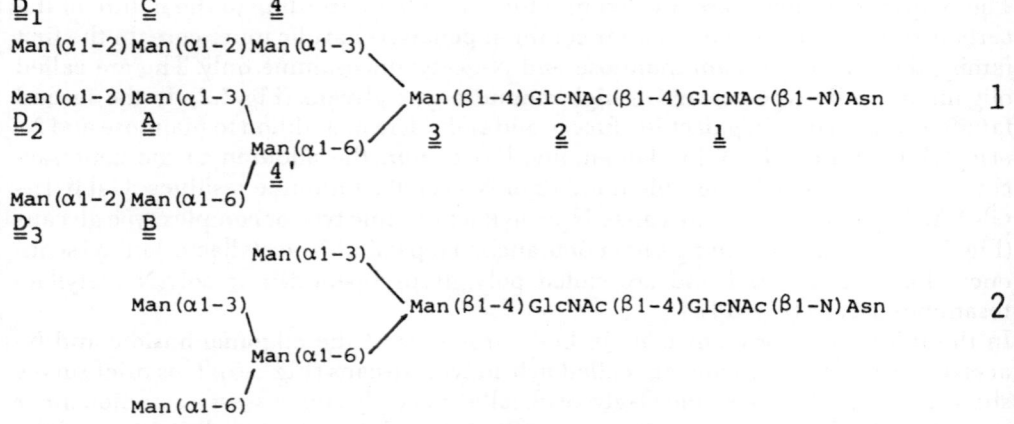

Fig. 3.2 Structure of oligomannosidic type (high-mannose type) glycans.
Structure 1 is present in calf thyroglobulin unit A, human IgD and myeloma IgM, Chinese hamster ovary cell glycoproteins, soybean agglutinin, Newcastle virus, scorpion hemocyanin. Structure 2 has been found in Taka-amylase, in hen ovalbumin and ovomucoid, in human myeloma IgM (for reviews, see Montreuil 1980, 1982, 1984 a, b)

52

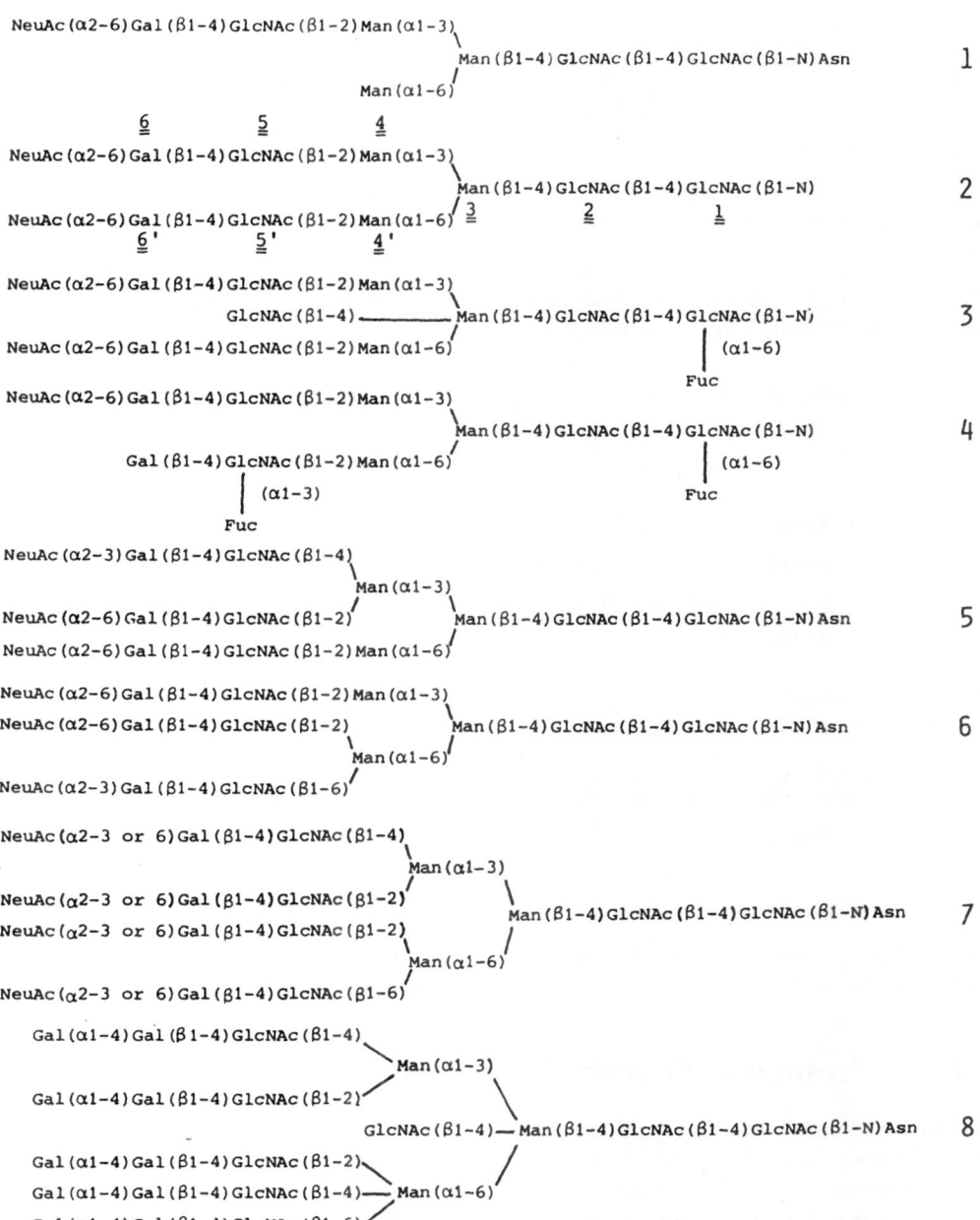

Fig. 3.3 Structure of glycans of the N-acetyllactosamine-type (complex type) glycans.
1 Monoantennary glycan of a secretory component from human milk and of human chorionic gonadotropin; 2 biantennary glycan of human serum transferrin; 3 monofucosylated and "bisected" (presence of a bisecting N-acetylglucosamine-residue) biantennary glycan of human IgG; 4 difucosylated biantennary glycan of human lactotransferrin; 5 triantennary glycan of human serum transferrin; 6 tri'-antennary glycan of human serum transferrin; 7 tetraantennary glycan of human α_1-acid glycoprotein; 8 pentaantennary glycan of turtle dove ovomucoid possessing β_1-serologic activity (for reviews, see Montreuil 1980, 1982, 1984 a, b)

$$R- \begin{bmatrix} \text{Gal}(\beta1-4)\text{GlcNAc}(\beta1-3) \end{bmatrix}_n \quad \text{Gal}(\beta1-4)\text{GlcNAc}(\beta1-2)\text{Man}(\beta1-3)$$

$$\text{Man}(\beta1-4)\text{GlcNAc}(\beta1-4)$$

$$\text{GlcNAc}(\beta1-N)\text{Asn}$$

$$R- \begin{bmatrix} \text{Gal}(\beta1-4)\text{GlcNAc}(\beta1-3) \end{bmatrix}_n \quad \text{Gal}(\beta1-4)\text{GlcNAc}(\beta1-2)\text{Man}(\beta1-6)$$

Fig. 3.4 General basic structure of biantennary poly(N-acetyllactosamine)-type glycans in which n varies from 1 to 15.

In human erythrocyte membrane, R = NeuAc(α 2–3)Cal(β 1–4)GlcNAc; NeuAc(α 2–6)Gal(β 1–4)GlcNAc, [Fuc(α 1–2)] Gal(β 1–4)GlcNAc (Group 0) or GalNAc (α 1–3) [Fuc(α 1–2)] Gal(β 1–4) GlcNAc (Group A) or Gal (α 1–3) [Fuc(α 1–2)] Gal (β 1–4) GlcNAc (Group B)

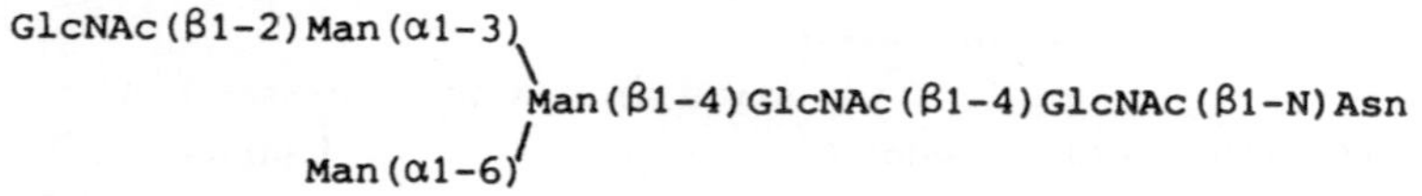

GlcNAc(β1–2)Man(α1–3)

Man(β1–4)GlcNAc(β1–4)GlcNAc(β1–N)Asn   **1**

Man(α1–6)

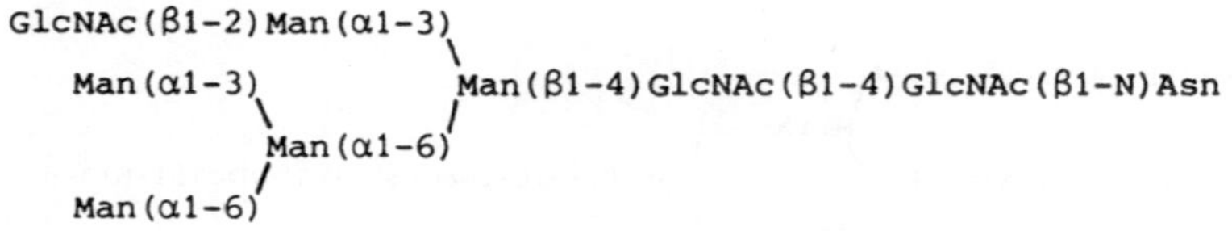

GlcNAc(β1–2)Man(α1–3)

Man(α1–3)

Man(β1–4)GlcNAc(β1–4)GlcNAc(β1–N)Asn   **2**

Man(α1–6)

Man(α1–6)

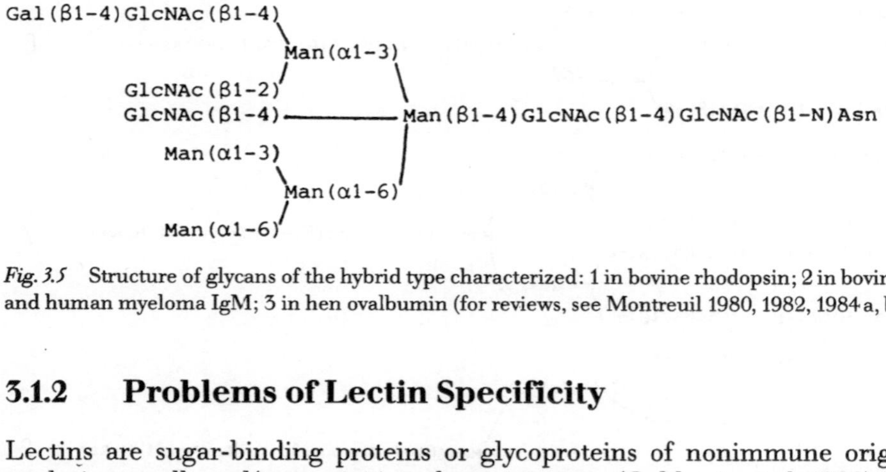

Gal(β1–4)GlcNAc(β1–4)

Man(α1–3)

GlcNAc(β1–2)

GlcNAc(β1–4)————Man(β1–4)GlcNAc(β1–4)GlcNAc(β1–N)Asn   **3**

Man(α1–3)

Man(α1–6)

Man(α1–6)

Fig. 3.5 Structure of glycans of the hybrid type characterized: 1 in bovine rhodopsin; 2 in bovine rhodopsin and human myeloma IgM; 3 in hen ovalbumin (for reviews, see Montreuil 1980, 1982, 1984a, b)

3.1.2   Problems of Lectin Specificity

Lectins are sugar-binding proteins or glycoproteins of nonimmune origin, which agglutinate cells and/or precipitate glycoconjugates (Goldstein et al. 1980). They have at least two sugar binding sites, the presence of which explains why lectins precipitate polysaccharides, glycoproteins and glycolipids and why they agglutinate cells. As these interactions can often be reversed by monosaccharides, immobilized lectins are now widely used to fractionate soluble and membrane glycoproteins of diverse origins, as well as glycopeptides or glycans derived from these glycoproteins (for reviews on lectins, see Lis and Sharon 1977, 1981, 1984, 1986a, b; Goldstein and Hayes 1978; Kornfeld and Kornfeld 1978; Goldstein and Poretz 1986).

However, very often, lectins are still classified according to the monosaccharide which inhibits the interaction between a lectin and a glycoprotein or which allows the specific

Table 3.1 Lectins commonly used for glycoprotein study and classified according to their monosaccharide specificity

α-D-Mannose, α-D-glucose	
Canavalia ensiformis	Con A
Lens culinaris	LCA
β-D-Galactose, N-acetyl-β-D-galactosamine	
Ricinus communis	RCA I
	RCA II
Glycine max (soybean)	SBA
Arachis hypogaea (peanut)	PNA
α-D-Galactose, N-Acetyl-α-D-galactosamine	
Griffonia simplicifolia I	GSA I
Dolichos biflorus	DBA
N-acetyl-β-D-glucosamine	
Triticum vulgare (Wheat germ)	WGA
α-L-Fucose	
Lotus tetragonolobus	LTA
Ulex europeus I	UEA I
α-N-acetylneuraminic acid	
Limulus polyphemus (limulin)	

elution of a bound glycoprotein from an immobilized lectin column. Table 3.1 presents such a classification of some commonly used lectins in terms of monosaccharide specificity. From such a classification, lectins such as concavanalin A (Con A) or *Lens culinaris* agglutinin (LCA) present the same specificity toward α-D-mannose or α-D-glucose. But, in fact, the interactions between lectins and glycoconjugates are much more complex and it is very important to have a better knowledge of the lectin specificity before using this lectin as a probe in the exploration of cell membrane glycoconjugates or as a tool to fractionate various glycoconjugates by affinity chromatography. This "complex" specificity of lectins can be determined by inhibition methods, in which complex oligosaccharides are tested for their ability to inhibit either hemagglutination (Debray et al. 1981) or precipitation of polysaccharides (Wu et al. 1988). Association constants can also be determined to study the interaction of a lectin with oligosaccharides (Baenziger and Fiete 1979a, 1979b; Ohyama et al. 1985). These studies show that only immobilized lectins with association constants greater than 4×10^6 M^{-1} for oligosaccharides in solution will bind these oligosaccharides.

A second approach is to study the affinity of the immobilized lectins toward complex oligosaccharides (Baenziger and Fiete 1979a; Debray and Montreuil 1981; Cummings and Kornfeld 1982a, b; Debray et al. 1983). From these studies, some conclusions can be drawn on the "true" sugar specificity of lectins.

3.1.2.1 Lectins Can Interact with Internal Sequences of an Oligosaccharide

Very few lectins bind to the terminal monosaccharides of a complex glycan, as it was believed for a long time, and most lectins are able to recognize and to bind to internal sequences of oligosaccharides. This first observation was made by Kornfeld and Ferris

(1975) who demonstrated that the most active part of an N-acetyllactosamine-type glycan recognized by Con A is the trimannosidic core substituted by two N-acetyl-glucosamine residues.

3.1.2.2 Lectins, Identical in Terms of Monosaccharide Specificity, Can Possess Different "Complex" Specificities

Some lectins, considered to be identical in terms of monosaccharide specificity, possess the ability to recognize fine differences in more complex oligosaccharidic structures. For instance, both Con A and LCA present the same specificity for α-D-mannose or α-D-glucose, but they show higher affinity for N-glycosylpeptides and related oligo-saccharides described in Fig. 3.6. Figure 3.6 also shows that the presence of a fucose residue enhances the affinity of LCA and is an important determinant (Debray et al. 1981; Kornfeld et al. 1981). Moreover, a given lectin is able to recognize very different saccharidic sequences. For instance, Con A presents a strong affinity either for oligo-mannosidic-type or for biantennary N-acetyllactosamine-type glycans (Fig. 3.6).

Different lectins are able to recognize different saccharidic sequences, but which belong to the same glycan structure. As these sequences are likely to be common to numerous glycoproteins, including cell membrane glycoproteins, some results, obtained with lectins during a fractionation of glycoproteins or in the study of cell surface glycans, have to be carefully interpreted.

3.1.2.3 Importance of the Spatial Conformation of Glycans in the Interaction with Lectins

For many lectins, the spatial conformation of glycans affects the interaction with the lectins (Montreuil et al. 1978, 1983; Montreuil 1980, 1982, 1983, 1984 a, b; Debray et al. 1981; Carver and Brisson 1984).

1. Lectin such as LCA presents a stronger affinity for glycopeptides than for glycans released from the N-glycosylpeptides by hydrazinolysis/N-reacetylation. This can be explained by the fact that the attachment of glycan to asparagine (glycoasparagines) or to peptides (N-glycosylpeptides) makes the trisaccharide sequence Man β-1,4-Glc NAc β-1,4-GlcNAc β-1,N-Asn rigid.

2. Many lectins, such as *Ricinus communis* agglutinin I (RCA I), present a higher affinity for saccharidic sequences substituted by sialic acid residues bound in the α-2,6 position of galactose rather than in the α-2,3 position. This can be related to the high flexibility of sialic acid residues around the α-2,6 linkage (Baenziger and Fiete 1979 b; Debray et al. 1981; Green et al. 1987 a).

   In contrast, for other lectins, such as isolectins E_4- or L_4-PHA (isolectins from *Phaseolus vulgaris* seeds with erythroagglutinating or leukostimulating activities), this higher mobility of sialic acid residues around the α-2,6 linkage can mask internal oligosaccharidic determinants recognized by the lectins. This can explain why sialylated glycans in the α-2,6 position of galactose are not recognized by the lectins while sialylated glycans, in the α-2,3 position of galactose will interact with them (Green and Baenziger 1987, Green et al. 1988).

56

57

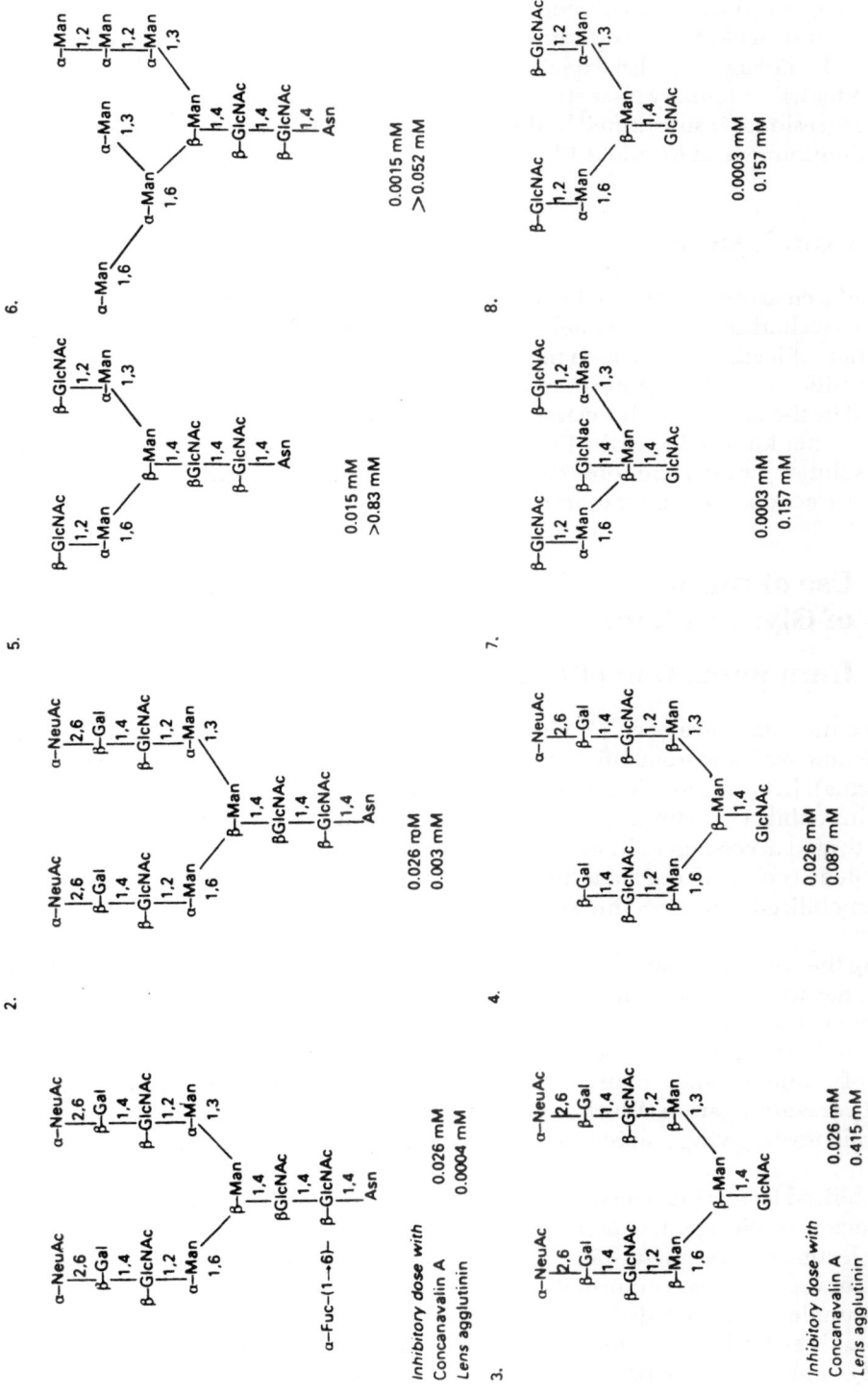

Fig. 3.6 Saccharidic determinants recognized by Concanavalin A and *Lens culinaris* agglutinin on glycans and oligosaccharides from glycoproteins with the N-glycosylamine linkage. Results are expressed as the minimum concentration of sugar (mM) necessary to inhibit completely one hemagglutinating dose (Debray et al. 1981)

3. The concept of considerable freedom of rotation in space of oligosaccharidic sequences around an α-1,6 glycosidic linkage as compared to the higher rigidity induced by the α-1,3 linkage can also explain the observed affinity of lectins such as LCA or L₄PHA for tri'-antennary N-acetyllactosamine-type glycans, in which one of the α-mannose residues is substituted by the third antenna at C-2 and C-6 (Kornfeld et al. 1981; Cummings and Kornfeld 1982a; Green et al. 1988).

3.1.3 Conclusions

These results clearly demonstrate the very precise specificity of lectins toward well defined oligosaccharidic sequences belonging either to N- or O-glycosylproteins. Thus, classification of lectins, based upon monosacchararide specificity must be substituted by new classifications taking into account this concept of "dominant oligosaccharides" recognized by the lectins. Such a classification has been proposed by Gallagher (1984). Moreover, better knowledge of the fine specificities of lectins is essential for their use, either in solution or after immobilization, in the study of cell membrane glycoconjugates or for fractionation of oligosaccharides.

3.2 Use of Immobilized Lectins for Fractionation of Glycoproteins

3.2.1 Immobilization of Lectins

Numerous methods have been proposed for preparing immobilized lectins, many of which are now available from different suppliers (Pharmacia, Miles, E.Y. Laboratories, I.B.F., Sigma). Lectins directly coupled to CNBr-activated agarose are the most popular ones. Immobilized lectins can be easily prepared with agarose, Sepharose 4B, or Ultrogel, activated according to March et al. (1974) or Kohn and Wilchek (1982), and at a coupling density of 2–10 mg lectin/ml settled gel. Immobilization and subsequent use of the immobilized lectins require some essential rules.

1. During the coupling reactions, lectin binding sites must be protected by addition of the monosaccharide specific to the lectin (competitive sugar).
2. In some cases, non specific (hydrophobic or ionic) adsorption of glycoproteins on the immobilized lectins has been observed. In order to avoid this phenomenon, new types of immobilization matrices of lectins have been developed, such as polyacrylic-hydrazido-agarose (Wilchek et al. 1984), on which lectins can be coupled with glutaraldehyde, giving a stable, non leaking, uncharged, and hydrophilic adsorbent.
3. Immobilized lectins must be stored at 4 °C in buffer containing 0.02 % sodium azide as a bacteriostatic agent, without loss of activity for several years.
4. Some lectins possess metal-binding sites and the presence of metal ions is important to induce a proper conformation of the lectin needed for carbohydrate binding. For example, Con A needs Mn^{2+} and Ca^{2+} for full activity (Bittiger and Schnebli 1976) and the buffers used for affinity chromatography on this lectin must contain $MnCl_2$ and $CaCl_2$ (1mM each). Sialic acid-binding lectins from *Limulus*

polyphemus (limulin or horseshoe crab agglutinin) as well as from *Carcinoscorpius rotunda cauda* (carcinoscorpin) are dependent on calcium ions for full activity (Robey and Liu 1981; Dorai et al. 1981).

5. In order to limit non specific interactions between glycoproteins and immobilized lectins, such as ionic interactions, the buffers must possess a moderate ionic strength (0.1–1 M in NaCl).

3.2.2 Use of Immobilized Lectins for Purification of Soluble Glycoproteins

Affinity chromatography on different immobilized lectins has been widely used to purify various soluble glycoproteins (for review, see Dulaney 1979; Lis and Sharon 1984, 1986 a, b). Some applications are shown in Table 3.2. We shall restrict discussion here to some particular points which could be useful during such fractionations.

Table 3.2   Soluble glycoproteins purified on immobilized lectins

Glycoprotein	Source	Lectin used	Reference
Major surface coat glycoprotein	*Trypanosoma brucei brucei*	LCA	Strickler et al. (1978)
Soluble glycoproteins	Marine sponge *Spongia officinalis*	Con A, WGA	Junqua et al. (1981)
Agglutinins of Leguminosae family	Leguminosae seeds	Con A	Bessler and Goldstein (1973)
Ovotransferrins	Hen egg	Con A	Iwase and Hotta (1977)
A, H blood group active substances	Hog gastric mucins	*Dolichos biflorus* agglutinin, LTA	Pereira and Kabat (1976)
Rat α_1-fetoprotein	Rat amniotic fluids	Con A	Bayard and Kerckaert (1981)
Acid phosphatase	Rat liver	Carcinoscorpin	Mohan et al. (1981)
Human chorionic gonadotropin	Human urines	Con A	Matsuura and Chen (1980)
Human immunoglobulin M (IgM)	Human plasma	RCAI	Harboe et al. (1975)

3.2.2.1 Use of Crossed Immunoaffinity Electrophoresis of Glycoproteins as a Guide to Lectin Affinity Chromatography

The interaction of plasma glycoproteins with lectins during electrophoresis was first described by Nakamura et al. (1960) and crossed immuno-affinoelectrophoresis (CIAE), introduced by Bøg-Hansen (1973, 1983) and Bøg-Hansen et al. (1977), combines the interaction between lectins and glycoproteins in the first dimension electrophoretic step with electrophoresis into an antibody-containing gel in the second dimension. This sensitive and powerful technique can give important information: (1) about the interaction between a given lectin and the glycoprotein to be purified, a good correlation being generally observed between the results obtained by affinity electrophoresis and by affinity chromatography on the immobilized lectin; and (2) about the microheterogeneity of carbohydrate moieties of the glycoproteins to be purified. However, parameters such as column binding capacity and elution conditions cannot be predicted from the affinity electrophoresis experiments.

3.2.2.2 Binding Capacity of the Immobilized Lectins and Problems of the Elution

Variations in the amount, as well as in the quality of the lectin immobilized on a gel, represent an important factor influencing the binding of glycoproteins. Affinity differences are often observed between different commercially available immobilized Con A preparations (Kerckaert and Bayard 1982). Similarly, affinity of WGA-Sepharose 4B for glycoproteins varies with the density of lectin coupled to the gel (Bhavanandan and Katlic 1979). This implies the calibration of the column before using a new batch of immobilized lectin with well known glycoproteins or glycopeptides. Three fractions are generally obtained, reflecting relative affinities of the immobilized lectins for the glycoproteins.

1. The nonreactive compounds are eluted at the void volume of the column with the equilibration buffer. This fraction, when the exact capacity of the immobilized lectin is not known, must be submitted to a new cycle of adsorption and elution on the same column, to insure that the immobilized lectin was not saturated during the first run.
2. The weakly reactive components give a fraction which is obtained by elution with the equilibration buffer. The separation of slightly differently interacting glycoproteins can even be obtained by using a long and thin column which is more efficient than a wider column containing the same volume of immobilized lectin. Ovotransferrin variants from hen egg white have been fractionated on a Con A-Sepharose column using this method (Iwase and Hotta 1977).
3. A strongly reactive fraction which is specifically desorbed by the addition of the appropriate sugar at a concentration between 0.1 and 0.5 M. According to the commercial origin of the immobilized lectin, a weakly reactive glycoprotein can be either bound and eluted with the lectin-reactive fraction or unbound and eluted with the lectin-nonreactive fraction.

The spatial conformations of the native glycoprotein or nonspecific hydrophobic interactions may modulate the accessibility of some glycans to the lectin, which can explain artifactual lectin, weakly reactive fractions. This inconvenience disappears when the glycoproteins are reduced and alkylated before fractionation on the immobilized lectin (Bayard et al. 1982).

Usually, immobilized lectin columns are eluted, after exhaustive washing with the equilibration buffer, with the same buffer but containing the competitive monosaccharide (0.01–0.5 M). However, when glycoproteins differing in their lectin-reactive oligosaccharide residues are fractionated, it may be interesting to elute the column with a concentration gradient of the competitive sugar. Fractionation of serum glycoproteins on a gradient-eluted column of Con A-Sepharose has been carried out by Baumstark (1983). Purified ovalbumins from hen egg white were also fractionated by Con A-Sepharose affinity chromatography into from major fractions by elution with a linear gradient of methyl α-D-glucoside from 0 to 0.2 M. The obtained fractions were related to the microheterogeneity of ovalbumin carbohydrate chains (Iwase et al. 1981). Immobilized *Carcinoscorpius rotunda cauda* agglutinin, a sialic acid-binding lectin, has been reported to be a useful tool for the resolution of sialoglycoproteins on the basis of their sialic acid content. Elution is obtained with a gradient of the lectin-specific disaccharide NeuAc α-2,6 GalNAc-itol (Mohan et al. 1981).

Sometimes the recovery of some bound glycoproteins is very low, even after elution with a high concentration of competitive sugar (0.5 M). This is due to either a very high affinity of the lectin for some saccharidic determinants of the glycoproteins or to multivalent interactions between the immobilized lectin and a very high proportion of the saccharide determinant on the glycoprotein as described (Bhavanandan and Katlic 1979). In the particular case of Con A, the recovery of high affinity glycoproteins can be improved by raising the temperature of the 0.5 M competitive sugar solution to 37 °C or even 60 °C. Nevertheless, this procedure cannot generally be applied since most lectins are less efficient at 37 °C than at room temperature (20 °C) or 4 °C (Lotan et al. 1977).

The specific displacement of glycoprotein by competitive sugars is reversible and after extensive washing with the equilibration buffer, the immobilized lectin can be used again with the same efficiency. However, stronger lectin-glycoprotein interactions can be displaced only by nonspecific desorption processes which very often cause the irreversible denaturation of the immobilized lectin. Such nonspecific displacements can be performed by pH change: elution of ovotransferrin from immobilized Con A was obtained at 4 °C with a 0.1 M acetic acid solution (Iwase and Hotta 1977); recovery of human chorionic gonadotropin from Con A-Sepharose has been obtained with 1 M acetic acid (pH 2.5) and the binding capacity of the column was restored by incubation either with Mn^{2+} or Mg^{2+} for 2 h. In contrast, elution with 1 M NH_4OH caused irreversible denaturation of the lectin (Matsuura and Chen 1980).

Elution of strongly bound glycoproteins from Con A-Sepharose can also be performed with 0.02–0.1 M borate buffers (pH 8.0), without denaturation of the lectin (Kennedy and Rosevear 1973; Junqua et al. 1981). High yield recovery can also be obtained by heating the Con A-Sepharose-glycoprotein complex for 3 min at 100 °C, in buffer containing 5 % (w/v) sodium dodecyl sulfate and 8 M urea, but with irreversible inactivation of the lectin (Poliquin and Shore 1980). Glycoproteins from *Trypanosoma congolense*, strongly bound to immobilized Con A, have been eluted by electrophoretic desorption (Reinwald et al. 1981).

3.2.2.3 Sequential Affinity Chromatography of Soluble Glycoproteins on Different Immobilized Lectins

The sequential use of immobilized lectins, with different and well-defined specificities toward saccharidic determinants, can be applied to fractionate complex mixtures of glycoproteins into classes, depending on their affinity for the different lectins. Immobilized Con A and WGA are the most currently used.

61

Generally, lectin affinity chromatography results in an enrichment of classes of different heterogeneous glycoproteins, possessing similar carbohydrate determinants recognized by the immobilized lectin and called "lectin receptors". But, on the other hand, sequential lectin affinity chromatography is a powerful method to unravel the microheterogeneity of glycans within a given glycoprotein family. For example, the immobilized sialic acid-binding lectin from *Carcinoscorpius rotunda cauda* has been used to resolve rat liver acid phosphatase into three different forms, according to their sialic acid content (Mohan et al. 1981). Eight ovalbumin subfractions have also been separated by successive lectin affinity chromatography on Con A- and WGA-Sepharose, according to their carbohydrate chains (Kato et al. 1984).

3.2.2.4 Development of High Performance Affinity Chromatography Using Immobilized Lectins

This new technique combines the high speed and resolution characteristics of high performance liquid chromatography (HPLC) with the selectivity of biospecific interactions (for review, see Larsson et al. 1983; Ernst-Cabrera and Wilchek 1987). However, very few applications of this technique, using lectins as bioselective ligands, have been published. Recently, a high performance Con A affinity column was used for the fractionation of human Con A nonreactive proteins: albumin, immunoglobulin G (IgG), transferrin, Gc-globulin, prealbumin and IgA. Con A reactive glycoproteins were eluted with 0.2 M methyl α-D-mannoside and contained immunoglobulins M (IgM), IgG, IgA, α-2-macroglobulin, transferrin, ceruloplasmin, hemopexin, prothrombin, and α1-antitrypsin as major components (Manabe et al. 1988).

3.2.3 Use of Immobilized Lectins for Fractionation of Membrane Glycoproteins

Lectin affinity chromatography is also widely used to fractionate integral membrane glycoproteins from various cell types (for reviews, see Lotan and Nicolson 1979; Hedo 1984). Some applications are given in Table 3.3.

The most important problem here is the stability of the immobilized lectins in the detergent solution used for membrane glycoprotein extraction. This detergent must be present during all further purification steps of the solubilized membrane proteins or glycoproteins. Since lectins are generally composed of subunits associated by noncovalent linkages, detergents may dissociate the lectin subunits and change their active conformation. Lotan et al. (1977) have systematically studied the effects of increasing concentrations of the most commonly used detergents on the glycoprotein binding efficiency of different immobilized lectins. Their results (see Table 3.4) show that the nonionic detergents, such as Triton X-100, are the most compatible with lectin affinity chromatography. Binding capacities of lectins are reduced in the presence of cationic, anionic, or zwitterionic detergents. The most efficient membrane solubilizing detergent, sodium dodecyl sulfate (SDS) releases subunits from the immobilized lectins and reduces their binding efficiencies. However, at a low concentration (0.08%), SDS has been successfully used to purify rat brain synaptosomal membrane glycoproteins by sequential affinity chromatography on immobilized Con A, *Ricinus communis* agglutinin, *Ulex europeus* agglutinin I, and WGA (Zanetta et al. 1981). Aono et al. (1985) have also reported that in the presence of 0.08 % SDS, the affinity chromatography of both soluble and membrane glycoproteins on Con A-Sepharose gives better results at 4 °C than at room temperature (26–28 °C). Recently, Scher et al. (1989) have shown that glutaraldehyde cross-linking of immobilized lectins (Con A, soybean agglutinin and WGA) does not reduce their binding capacities for membrane glycoproteins. This procedure allows the subsequent elution of bound glycoproteins with 0.5 % SDS, and without coelution of lectin subunits.

Table 3.3 Membrane glycoproteins purified on immobilized lectins

Glycoprotein	Source	Lectin used	Reference
Virus membrane glycoproteins	Different virus	LCA	Hayman et al. (1973)
Membrane glycoproteins	Pig lymphocyte plasma membranes	LCA	Hayman and Crumpton (1972)
Major glycoprotein	Milk fat globule membranes	Con A	Imam et al. (1981)
Membrane glycoproteins	Human blood platelets	LCA, WGA	Clemetson et al. (1977)
Membrane glycoproteins	Novikoff tumor cells	RCA, SBA	
		Con A, WGA	Gleeney and Walborg (1979)
Major sialoglycoprotein	Human lymphoblastoid cells	WGA, RCA, Con A	Saito et al. (1978)
Membrane receptors for peanut lectin	Human erythrocytes	PNA	Carter and Sharon (1977)
Glycophorin	Human erythrocytes	WGA	Kahane et al. (1976)
Membrane glycoproteins	Rat brain synaptic vesicles	Con A, RCA, WGA, UEAI	Zanetta et al. (1981)

Table 3.4 Effects of commonly used detergents on the binding efficiency of immobilized lectins according to Lotan et al. (1977)

Detergent concentration (%)	Glycoprotein-binding efficiency of immobilized lectin (%)		
	Ricinus communis agglutinin	Concanavalin A	Wheat germ agglutinin
Triton X-100			
0.1	100	85	90
1.0	100	76	76
2.5	100	74	79
Desoxycholate			
0.1	100	60	85
1.0	100	46	39
2.5	100	20	10
Sodium dodecylsulfate			
0.05	100	70	89
0.1	70	26	13

3.2.4 Limitations of Lectin Affinity Chromatography for Purification of Soluble and Membrane Glycoproteins

Lectin affinity chromatography of complex mixtures of soluble or membrane glycoproteins very rarely permits the purification of a specific glycoprotein in a single step. However, affinity chromatography on immobilized WGA allowed the almost quantitative purification of glycophorin, the major human erythrocyte membrane glycoprotein (Kahane et al. 1976) as well as insulin receptor purification (Hedo et al. 1981) in a single step. Laminin, a major component of the basement membrane, has been purified from extracts of EHS sarcoma cells by affinity chromatography on immobilized *Griffonia simplicifolia* I-B$_4$ lectin (Shibata et al. 1982).

A new mannose-specific plant lectin isolated from the snowdrop bulb *Galanthus nivalis* was recently shown to be very useful, after immobilization on Sepharose 4B, for the one-step purification of IgM-type antibodies from murine sera and represents a very potent tool for the purification of IgM-type monoclonal antibodies (Shibuya et al. 1988 a). These glycoproteins, containing oligomannosidic-type glycans recognized by the lectin, bind tightly to the immobilized lectin and are eluted with 0.1 M methyl α-D-mannoside. In contrast, human IgM which also contains oligomannosidic type glycans is not retained on the lectin column and α-2-macroglobulin is the only glycoprotein from human sera which was bound to the immobilized lectin. An α-D-galactose specific lectin from Jackfruit seed *Artocarpus integrifolia*, Jacalin (Azevedo Moreira and Ainouz 1981) represents another powerful tool for the affinity purification of human IgA$_1$ and IgD (Roque-Barreira and Campos-Neto 1985; Kondoh et al. 1987; Aucouturier et al. 1987; Zehr and Litwin 1987). Studies of saccharide binding to the lectin (Sastry et al. 1986) had revealed a specific recognition of the core disaccharide of mucin-type glycoproteins Gal β-1,3 GalNAc, which could explain the strong reactivity of Jacalin with human IgA$_1$ and IgD, which possess several 0-glycosidically linked Gal β-1,3 GalNAc determinants in their hinge region (Baenziger and Kornfeld 1974; Mellis and Baenziger 1983 b).

Neither IgM nor any of the IgG subclasses bind to the immobilized lectin. However, the IgA$_2$ reactivity of Jacalin is still a matter of controversy (Aucouturier et al. 1988; Kobayashi et al. 1988), which may depend on the origins (countries and cultivated variants) of the different Jacalins used in these studies.

However, lectin chromatography of glycoproteins, from complex biological extracts, usually results in an enrichment of different classes of heterogeneous glycoproteins, which possess similar carbohydrate determinants recognized by the immobilized lectin. As different lectins are able to recognize different saccharidic sequences belonging to the same glycan and as these glycans are likely to be common to numerous glycoproteins, the different lectins will react in fact with a broad spectrum of glycoproteins.

However, when a glycoprotein, in a first step, can be purified from other contaminating glycoproteins, either by conventional biochemical procedures or by immunoaffinity chromatography, lectin affinity chromatography will be, in a second step, a very powerful tool to determine the glycan microheterogeneity within this glycoprotein family.

A practical approach in the use of immobilized lectins to fractionate glycoproteins (immobilization of lectins, affinity chromatography of glycoproteins on immobilized Con A and WGA) is described by Montreuil et al. (1986).

3.3 Use of Immobilized Lectins for Fractionation of Glycopeptides and Oligosaccharides

3.3.1 Introduction

Although the purification and fractionation of glycoproteins by lectin affinity chromatography present some limitations, the procedure represents a powerful tool for the fractionation of glycopeptides and glycans. This is essentially due to the fact that nonspecific hydrophobic or ion-exchange interactions often observed between glyco-ionicproteins and lectins are very rare in the case of glycans and glycopeptides. These compounds interact specifically with the lectins and can be fractionated on the basis of an "actual" affinity chromatography.

Consequently this implies, even more than in glycoprotein fractionations, a very precise knowledge of the exact specificities of the lectins to be used. Thus, once the precise specificity of an immobilized lectin is well defined, it becomes possible to predict the primary structure of bound glycopeptides or oligosaccharides. On this basis, schemes of fractionation of glycopeptides or glycans, obtained from soluble or membrane-bound glycoproteins, by lectin affinity chromatography have been proposed (for reviews, see Finne et al. 1980; Montreuil et al. 1986; Merkle and Cummings 1987a; Osawa and Tsuji 1987).

3.3.2 Fractionation of N-Glycosylpeptides Using Immobilized Lectins with Affinity for Internal Saccharidic Sequences

3.3.2.1 Affinity Chromatography on Immobilized Concanavalin A

The first group fractionation of glycopeptides on immobilized Con A was proposed by Ogata et al. (1975). From the results of binding of ^{3}H-labeled glycopeptides and oligosaccharides to a Con A-Sepharose column, they showed that at least two unsubstituted or 2-0-substituted α-mannose residues were necessary for the glycans to bind on the column.

On a Con A-Sepharose column, a mixture of N-glycosylpeptides and/or related oligosaccharides can be fractionated into four classes as shown in Fig. 3.7 (Debray and Montreuil 1981; Debray et al. 1983; Debray et al. unpubl. data):

1. Nonbound glycopeptides (or oligosaccharides), are eluted at the void volume of this column with the equilibration buffer.
2. A retarded fraction (RF) can be obtained by elution with the equilibration buffer, reflecting a weak affinity between the lectin and the glycans.
3. A weakly bound fraction (WBF) is retained on the Con A-Sepharose column and is eluted with a low concentration (0.01 M) of methyl α-D-glucoside.
4. A strongly bound fraction (SBF) is eluted with a high concentration (0.3 M) of methyl α-D-glucoside.

Briefly, the nonbound fraction (NBF) may contain tetra- or triantennary N-acetyllactosamine-type glycopeptides, or biantennary structures with a "bisecting" GlcNAc residue on the β-linked mannose residue.

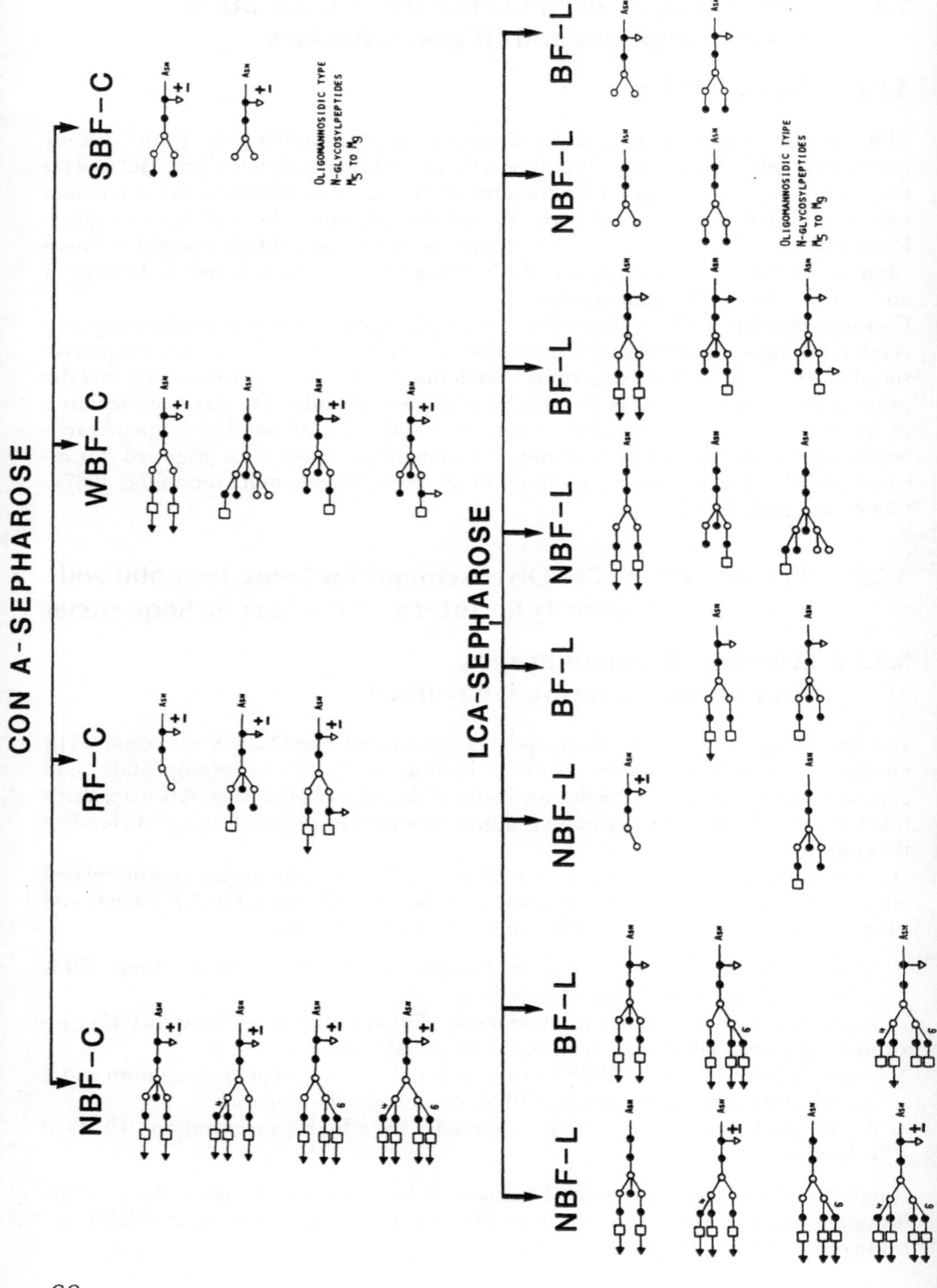

66

The retarded fraction (RF) may contain N-glycosylpeptides with terminal nonreducing galactose residues with either a "bisecting" GlcNAc residue or a fucose residue substituting GlcNAc residue(s) in C-3 position or both, giving steric hindrances that weaken the interaction with immobilized Con A. Biantennary N-glycosylpeptides are weakly bound and are eluted with 0.01 M methyl α-D-glucoside (WBF). It is noteworthy that some biantennary structures with a "bisecting" GlcNAc residue are weakly retained on the condition that at least one unsubstituted GlcNAc β-1,2-Manα-1,3 sequence is available on the glycan. Some hybrid-type glycopeptides are also weakly bound by immobilized Con A, but their elution volume is larger.

Oligomannosidic-type glycopeptides are strongly bound on Con A-Sepharose and are eluted with a high concentration (0.3 M) of methyl α-D-glucoside, giving a broad and trailing profile (SBF). Immobilized Con A also strongly reacts with biantennary glycopeptides which possess two terminal nonreducing GlcNAc residues.

These results confirm and extend previous data obtained by different groups (Ogata et al. 1975; Krusius et al. 1976; Narasimhan et al. 1979, 1986; Baenziger and Fiete 1979 a). However, small differences in resolution on immobilized Con A can be found, which may depend on batch-to-batch variations of commercial Con A-Sepharose or on the chromatographic conditions (buffers, concentration of eluting methyl α-D-glucoside solutions). For example, depending on the amount of immobilized Con A/ml of Sepharose, weakly retained biantennary N-acetyllactosamine-type glycopeptides can be eluted as a retarded fraction with the equilibration buffer, free of competitive sugar.

The degree of separation of some glycopeptides may also depend on the geometry of the Con A column: long, thin columns can be more efficient than wider columns containing the same amount of immobilized lectin.

Before using a new batch of immobilized Con A, calibration of the immobilized lectin with radiolabeled glycopeptides or glycans of known structure must be performed. These glycopeptides or glycans can be prepared from commercially available glycoproteins such as human α1-acid glycoprotein, human serum and lactotransferrin, hen ovalbumin as described (Finne and Krusius 1982; Montreuil et al. 1986).

Glycopeptides can be N-[14C] or [3H]-acetylated according to Koide et al. (1974), to easily follow the fractionation. Oligosaccharides, released from glycoproteins or glycopeptides either by chemical hydrazinolysis, or by enzymatic cleavage with various endoβ-N-acetylglucosaminidases or aspartyl-N-acetylglucosaminidase peptide-N-glycosidases, can be labeled at the reducing-terminal GlcNAc residues by reduction with [3H]-potassium borohydride according to Takasaki and Kobata (1974).

Since intensive use may lead to a decrease in the binding capacity of the immobilized Con A due to lectin leakage, the calibration procedure with standard oligosaccharides must be periodically performed. When the exact binding capacity of the immobilized lectin is not known, the unretained fraction must be recycled on the regenerated column. It is noteworthy that N-glycosylpeptides present the same affinity and behavior on an immobilized Con A column as glycans released from them, either by chemical or enzymatic cleavages.

Fig. 3.7 Interaction of glycopeptides with the N-glycosylamine linkage with immobilized Concanavalin A and *Lens culinaris* agglutinin.
● GlcNAc, ○ Man, □ Gal, ↓ Fuc, ← NeuAc (Debray and Montreuil 1981; Kornfeld et al. 1981)

3.3.2.2 Affinity Chromatography on Immobilized Lens culinaris Agglutinin

As shown in Fig. 3.7, each glycopeptide family obtained from Con A Sepharose can be further divided into two populations by affinity chromatography on immobilized *Lens culinaris* agglutinin (LCA), on the basis of the presence of a fucose residue α-1,6-linked to the N-acetylglucosamine residue involved in the N-glycosylamine bond.

As previously reported (Debray and Montreuil 1981; Debray et al. 1981; Kornfeld et al. 1981), immobilized LCA interacts with biantennary glycopeptides or with triantennary glycopeptides possessing an L-fucose residue at the C-6 position of the GlcNAc residue linked to the asparagine residue. However, an important condition for the latter interaction to take place is that the third antennae must substitute the α-1,6-linked mannose of the inner core in the C-6 position. If this substitution is at the C-4 position of the α-1,3-linked mannose, immobilized LCA does not present any affinity for this triantennary glycopeptide (Kornfeld et al. 1981). However, it is surprising that some monosialylated tetraantennary glycopeptides interact with the lectin (Samor et al. 1986). This may be correlated with the presence of a single sialic acid residue, which provides the glycopeptide with a conformation that allows the lectin to recognize its specific determinant.

LCA, in contrast to Con A, interacts with biantennary structures possessing a "bisecting" GlcNAc residue, but on the condition that these structures possess an L-fucose residue in the C-6 position of the GlcNAc residue involved in the N-glycosylamine linkage.

The binding of biantennary glycopeptides to LCA-Sepharose is enhanced by the exposure of terminal GlcNAc residues, but the "limit" glycan recognized by the lectin must possess two α-D-mannose residues. Fractionation with immobilized LCA, as well as with related *Vicia faba* (VFA) and *Pisum sativum* (PSA) agglutinins, can be applied only to N-glycosylpeptides. Reduced glycans, released from N-glycosylpeptides by hydrazinolysis and still possessing the α-1,6-fucose determinant, are no longer retarded on immobilized VFA or PSA or bound on immobilized LCA (Debray and Montreuil 1981; Yamamoto et al. 1982: Katagiri et al. 1984). This is an important point to consider in the fractionation procedure to be used, since the discriminating power of an immobilized LCA-Sepharose column is considerable and cannot be neglected. The specificity of immobilized VFA or PSA is similar to that of immobilized LCA. However, their affinity seems to be lower than that of LCA, since glycopeptides firmly bound on LCA and eluted with 0.15 M methyl α-D-glucoside solution are only retarded on immobilized VFA or PSA (Debray et al. 1981; Debray and Montreuil 1981; Kornfeld et al. 1981; Yamamoto et al. 1982; Katagiri et al. 1984).

All these results show that four lectins (Con A, LCA, VFA and PSA), considered for a long time to be identical in terms of α-D-mannose or glucose specificity, possess the ability to recognize fine differences in more complex carbohydrate structures. While immobilized Con A represents a very useful tool for the fractionation of N-acetyllactosamine-, oligomannosidic-, or hybrid-type glycopeptides or related oligosaccharides, the three other lectins are very useful for the separation of some N-acetyllactosamine-type glycopeptides on the condition that an L-fucose residue substitutes in C-6 the GlcNAc residue involved in the N-glycosylamine bond.

As shown in Fig. 3.7, sequential affinity chromatography on immobilized Con A and LCA allows a very precise fractionation of complex mixtures of N-glycosylpeptides,

68

prepared either from soluble or membrane-bound glycoproteins by proteolytic digestion with pronase into eight families. This procedure also permits a precise identification of minute quantities of glycopeptides on the basis of the comparison of their elution profiles with those of known oligosaccharidic structures.

3.3.2.3 Affinity Chromatography on Other Lectins with a Specificity for Internal Saccharidic Sequences

As originally proposed by Cummings and Kornfeld (1982b), other immobilized lectins
able to recognize different internal sequences of glycans can be serially used to sub-
fractionate N-glycosylpeptide families obtained after affinity chromatography on immobilized Con A and LCA/or PSA.

Immobilized Phaseolus vulgaris Isolectin E_4 (E_4-PHA)

Phaseolus vulgaris seeds contain five tetrameric isolectins (L_4, L_3E_1, L_2E_2, L_1E_3 and
E_4). The L_4 isolectin (L_4-PHA) or leukoagglutinating phytohemagglutinin binds to
lymphocytes and is strongly mitogenic for these cells. In contrast, the E_4 isolectin (E_4-
PHA) or erythroagglutinating phytohemagglutinin agglutinates erythrocytes, but not
lymphocytes (for a review, see Goldstein and Poretz 1986). The lectins from *Phaseolus
vulgaris* are not inhibited by simple sugars except N-acetylgalactosamine at rather
high concentration. Kornfeld and Kornfeld (1970) had reported that the binding site for
E_4-PHA on human erythrocytes was an N-acetyllactosamine-type glycopeptide and
that the galactose residues were important determinants for binding.
Elucidation of the fine specificity of E_4-PHA was possible by the analysis of the elution
behavior of N-glycosylpeptides and related oligosaccharides on columns of immobilized E_4-PHA.
According to this method, Irimura et al. (1981) showed that a biantennary N-acetyllactosamine-type reduced glycan, with a "bisecting" GlcNAc residue and two terminal
nonreducing galactose residues, isolated from human glycophorin A, was retarded on
an E_4-PHA column. Interaction with the lectin was abolished after removal of the terminal galactose residues. Later, Cummings and Kornfeld (1982a) showed that the presence of a "bisecting" GlcNAc residue was an important determinant for high affinity
binding to E_4-PHA-Sepharose.
Studies from Yamashita et al. (1983) with "bisected" glycans and Narasimhan et al.
(1986) with "bisected" N-glycosylpeptides, show that the peptidic part or the presence
of an α-1,6-linked fucose at the innermost GlcNAc residue of the core does not influence the binding to E_4-PHA. Their results also indicate that the Galβ-1,4-GlcNAcβ-
1,2-Man α-1,6 branch is an important determinant in the lectin-oligosaccharide interaction which must not be sialylated by α-2,6-linked sialic acid residue. In contrast,
Green and Baenziger (1987) have shown that the presence of α-2,3-linked sialic acid
residues on N-glycanase-released oligosaccharides enhances interaction of these
oligosaccharides with E_4-PHA. The presence of a GlcNAc residue β-1,2-linked to the
core Man α-1,3 residue was also found to be essential for binding, but sialylation of the
Galβ-1,4-GlcNAcβ-1,2-Man α-1,3 branch does not interfere with binding. Yamashita
et al. (1983) also found that the addition of a third Galβ-1,4-GlcNAcβ-1,4 antenna on

69

the core α-1,3-Man residue did not alter the binding of a "bisected" triantennary oligosaccharide with E$_4$-PHA. However, substitution of the Man α-1,6 residue by another Galβ-1,4-GlcNAc sequence completely abolished the binding.

It was also found that substitution of the outer branch GlcNAc residues with α-1,3-linked fucose residues decreased the affinity of E$_4$-PHA for N-glycosylpeptides (Santer et al. 1983).

Differences in the resolution of N-glycosylpeptides or oligosaccharides on immobilized E$_4$-PHA have been reported and depend on the amount of immobilized lectin per milliliter of gel. At a low coupling density of lectin (less than 1 mg/ml of gel), "bisected" biantennary glycopeptides are eluted as a retarded fraction with the equilibration buffer, and in that case, a better resolution of glycopeptides is obtained with a long column of the immobilized lectin. But, at higher coupling densities (3 mg/ml of gel), glycopeptides interact strongly with the lectin and must be eluted with a 0.4 M N-acetylgalactosamine solution or with 0.05 M glycine-HCl buffer, pH 3.5 (Mellis and Baenziger 1983 a).

Moreover, in a recent study, Green and Baenziger (1987) showed that the specificity of immobilized E$_4$-PHA for free, reduced oligosaccharides released from N-glycosylproteins by an aspartyl-N-acetylglucosaminidase, peptide-N-glycosidase (N-glycanase) differs significantly from that established with N-glycosylpeptides. If desialylated "bisected" biantennary oligosaccharides still presents the greater interaction with the immobilized lectin, desialylated "nonbisected" oligosaccharides containing one, two, three, or four N-acetyllactosamine antennae are retarded to varying extents by E$_4$-PHA, in contrast to N-glycosylpeptides containing these glycans which do not interact with the lectin. This interaction is decreased or abolished by removal of terminal nonreducing galactose residues. This phenomenon may be related to the different spatial conformations of free glycans and of N-glycosylpeptides with the same oligosaccharidic structures (see sect. 3.1.2.3).

In conclusion, the specificity of E$_4$-PHA is very complex and, as first pointed out by Yamashita et al. (1983) and then by Green and Baenziger (1987), application of an E$_4$-PHA column for the fractionation and analysis of N-glycosylpeptides or related nonreduced or reduced oligosaccharides, released by hydrazinolysis or N-glycanase cleavages, must be handled with care. Before use, E$_4$-PHA columns must therefore be carefully calibrated with standard oligosaccharides (free and nonreduced, or free and reduced oligosaccharides, or N-glycosylpeptides) related to the oligosaccharides to be fractionated.

Immobilized Phaseolus vulgaris Isolectin L$_4$(L$_4$-PHA)

The carbohydrate-binding specificity of the leukoagglutinating phytohemagglutinin L$_4$-PHA was first examined by inhibition of precipitation of carcinoembryonic antigen with N-glycosylpeptides and oligosaccharides of known structures (Hammarström et al. 1982). The disaccharide GlcNAcβ-1,2-Man was shown to be the minimal structural unit required for L$_4$-PHA binding, whereas the most effective structure appeared to be the pentasaccharide Galβ-1,4-GlcNAcβ-1,2 [Galβ-1,4-GlcNAcβ-1,6] Man.

Later, Cummings and Kornfeld (1982a) studied the binding of different N-glycosylpeptides to immobilized L$_4$-PHA. These authors established that tri'- and tetraantennary N-acetyllactosamine-type glycopeptides containing the above pentasaccharidic sequence were specifically retarded; bi- and triantennary glycopeptides lacking the

70

Galβ-1,4-GlcNAcβ-1,6 branch as well as "bisected" biantennary glycopeptides showed no interaction with the lectin. The interaction of reacting tri'- and tetraantennary glycopeptides was completely abolished by the removal of the galactose residues, but the presence of an α-1,6-linked fucose at the innermost GlcNAc of the core did not influence the behavior of the glycopeptides on a L_4-PHA column.

More recently, Green and Baenziger (1987) showed that the immobilized lectin interacted with free, reduced asialotri'- and -tetraantennary oligosaccharides released by N-glycanase, but also with asialobi- and triantennary as well as with "bisected" asialobiantennary oligosaccharides. All these glycans are retarded but to different extents on a long column of L_4-PHA (40 cm × 0.35 cm i. d.). The presence of α-2,6-linked NeuAc residues has an inhibitory effect on the interaction of the oligosaccharides with the lectin; in contrast, the presence of α-2,3-linked NeuAc residues significantly enhances the binding of these sialylated oligosaccharides.

It is also noteworthy that in this study, Green and Baenziger found that both immobilized L_4-PHA and E_4-PHA present similar specificities for all these free, reduced oligosaccharides. Later, using N-glycanase-released N-linked oligosaccharides of known structure from bovine fetuin, Green et al. (1988) confirmed that the antenna arising from the α-1,6-linked core mannose residue played the predominant role in interaction with L_4-PHA. All the triantennary glycans with a terminal α-2,3-linked NeuAc residue on this antenna interact strongly with the lectin while those with an α-2,6-linked NeuAc residue on this branch do not interact or interact weakly with the lectin.

More recently, Bierhuizen et al. (1988) analyzed the effects of branching and of substitution of antennae by sialic acid and fucose residues on the interaction of N-glycosylpeptides and related oligosaccharides with immobilized L_4-PHA. They found that asialotri'- and tetraantennary glycans were strongly retarded whereas asialobi- and triantennary glycans lacking the Galβ-1,4-GlcNAc-β-1,6 antenna were only weakly retarded. The interaction with the lectin was completely abolished when either α-2,6-linked NeuAc or α-1,3-linked fucose substituted the galactose or the N-acetylglucosamine residue of the Galβ-1,4-GlcNAcβ-1,2-Man α-1,6 antennae. The same substitutions on the Galβ-1,4-GlcNAcβ-1,6-Man α-1,6 antenna decreased but did not abolish the affinity of the lectin for the glycans.

This study draws attention to the necessity to desialylate and to defucosylate the glycans before fractionation on immobilized L_4-PHA, to obtain a better resolution of interacting N-glycosylpeptides and related oligosaccharides.

In conclusion, application of a L_4-PHA column for the fractionation and analysis of N-glycosylpeptides or related non-reduced, or reduced oligosaccharides, released by hydrazinolysis or N-glycanase treatment must be also handled with care. As previously emphasized for E_4-PHA, the immobilized L_4-PHA column to be used must be calibrated with the appropriate standards prepared in a similar way as the oligosaccharidic structures to be analyzed.

Immobilized Wheat Germ Agglutinin (WGA)

The carbohydrate-binding specificity of WGA has been studied by different techniques (for review, see Goldstein and Poretz 1986). Inhibition by various oligosaccharides of WGA-induced hemagglutination had shown that N,N',N''-triacetylchitotriose was 10-fold more inhibitory than N,N'-diacetylchitobiose and 100-fold more than N-acetylglucosamine (Debray et al. 1981). These results were in agreement with those of Allen et al. (1973), Goldstein et al. (1975) and Monsigny et al. (1978), showing that the combining site of WGA was complementary to a sequence of three β-1,4-linked N-acetylglucosamine residues. However, WGA is also able to interact specifically with nonreducing terminal N-acetylneuraminic acid residues of glycoproteins and glycolipids because of the similar configuration due to the acetamido group of N-acetylneuraminic acid (Bhavanandan and Katlic 1979; Peters et al. 1979; Monsigny et al. 1980). These interactions also involve a charge effect and what Monsigny et al. (1980) call an "avidity effect": glycoconjugates (glycoproteins, glycolipids or glycopeptides) with a high density of terminal nonreducing N-acetylglucosamine or N-acetylneuraminic acid residues will interact with WGA.

Later, Yamamoto et al. (1981) and Debray et al. (1983) studied the structural requirements for the interaction of N-glycosylpeptides and related oligosaccharides on immobilized WGA-Sepharose columns with a low coupling density of lectin (2–5 mg/ml of gel).

"Bisected" hybrid-type N-glycosylpeptides isolated from ovalbumin bind to the column and sialylated or asialo-biantennary "bisected" glycopeptides are retarded, whereas oligomannosidic-type, "nonbisected" biantennary or tri- and tetraantennary glycopeptides do not interact with the immobilized WGA.

The GlcNAcβ-1,4-Manβ-1,4-GlcNAcβ-1,4-GlcNAcβ-1,N-Asn sequence represents the structural determinant recognized by WGA. The inner N,N'-diacetylchitobiose-Asn core is an important determinant for binding, since "bisected" glycans, released from N-glycosylpeptides by hydrazinolysis, are no longer retarded or bound on an immobilized WGA column. Substitution of the innermost GlcNAc residue of the core at C-6 by a fucose residue prevents the binding of a "bisected" biantennary N-acetyllactosamine-type glycopeptide to a WGA column, probably due to steric hindrance.

Poly(N-acetyllactosamine)-type glycans bind with high affinity to immobilized WGA and internal GlcNAc residues, but not the external NeuAc residues, represent the high affinity binding sites (Gallagher et al. 1985). Ivatt et al. (1986a) confirmed that linear or branched poly(N-acetyllactosamine)-type glycopeptides prepared from Band 3 glycoprotein of fetal or adult erythrocytes were tightly bound to immobilized WGA, possibly through GlcNAc β-1,3-Galβ-1,4-GlcNAcβ-1,3-Gal sequences either at the nonreducing and/or within the poly(N-acetyllactosamine)chains. They also showed that the interaction of linear poly(N-acetyllactosamine)-type glycans with WGA was very complex; the presence of sialic acid residues decreases the affinity of poly(N-acetyllactosamine)-type glycans for WGA. In contrast, the presence of α-1,3-linked fucose residues on GlcNAc β-1,3-Gal β-1,4-GlcNAc β-1,3-Gal repeating units seems to be essential in the interaction of some poly(N-acetyllactosamine)-type glycans with immobilized WGA (Ivatt et al. 1986b). They also showed the role of lectin coupling density: at high WGA density (10 mg/ml of gel), a higher affinity interaction was demonstrated between WGA and poly(N-acetyllactosamine)-type glycans.

Recently, Renkonen et al. (1988) investigated the behavior of oligosaccharides derived from poly(N-acetyllactosamine)-type glycans of teratocarcinoma cells with immobilized WGA. They found that at low coupling density (1.9 mg/ml of gel), a separation of five oligosaccharides could be obtained; Gal β-1,4-GlcNAc, GlcNAc β-1,3-Gal and GlcNA β-1,3-Gal β-1,4-GlcNAc are only weakly retarded, but GlcNAc β-1,6-Gal β-1,4-GlcNAc are more strongly retarded and Glc NAc β-1,6-Gal binds to the WGA column and must be eluted with a 0.2 M GlcNAc solution. The importance of WGA coupling density was also emphasized by Tarrago et al. (1988). They showed that at high WGA density (25 mg/ml of gel), a reduced sialyllacto-N-neohexose with an α-2,6-linked NeuAc residue on the terminal galactose isolated from human milk (Fig. 3.8) interacted weakly with immobilized WGA and could be isolated from all other monosialyloligosaccharides present in human milk. No interaction was observed at a lower WGA density; the sialic acid residue is not involved in the interaction with WGA and even interfered with binding since the asialooligosaccharide is more retarded on the high density WGA column. For these authors, the weak affinity of WGA for the oligosaccharide is probably due to the presence of two GlcNAc residues unsubstituted at the C-3 hydroxyls, since the tetrasaccharide derived from the neutral hexose by β-galactosidase digestion showed the same behavior on the WGA column and the trisaccharide GlcNAc β-1,3-Gal β-1,4-Glc-ol did not interacted at all.

```
NeuAc(α2-6)Gal(β1-4)GlcNAc(β1-3)
                                 \
                                  Gal(β1-4)Glc
                                 /
     Gal(β1-4)GlcNAc(β1-6)
```

Fig. 3.8 Structure of a monosialylhexasaccharide isolated from human milk by WGA affinity chromatography (Tarrago et al. 1988)

As WGA displays a higher affinity for the α-2,3-rather than for the α-2,6-isomer of N-acetylneuraminyllactose (Kronis and Carver 1982), Tarrago et al. (1988) suggested that the separation of sialyloligosaccharides or of weakly interacting ligands (association constants $<10^4 M^{-1}$) should be possible using high-performance liquid affinity chromatography at a high concentration of immobilized WGA (>100 mg/ml of silica). Immobilized WGA also interacts with glycopeptides presenting a high density of 0-glycosidically-linked sialyloligosaccharides (Bhavanandan et al. 1977; Furukawa et al. 1986).

In summary, the interaction of immobilized WGA with N-glycosidically-linked glycans is very complex and depends on the amount of immobilized lectin as well as on the length and size of the columns used for fractionation. As previously pointed out for E_4- and L_4-PHA, the immobilized WGA columns to be used must be carefully calibrated with the appropriate standards prepared in a similar way as the oligosaccharidic structures to be analyzed.

Immobilized Oriza sativa Agglutinin (Rice Lectin)

Recently, an immobilized lectin isolated from rice seeds (Poola et al. 1986) was proposed as an alternative to E_4-PHA and WGA for the fractionation of "bisected" biantennary N-glycosylpeptides by Poola and Narasimhan (1988). In contrast to WGA, N, N'-diacetylchitobiosyl-Asn was found to be the minimum structure binding to rice lectin. A bisecting β-1,4-linked GlcNAc attached to β-linked core mannose enhances the binding of sialo-, asialo-, or asialogalacto-N-acetyllactosamine-type biantennary N-glycosylpeptides, but the presence of one or two α-2,6-linked NeuAc residues decreases the binding capacity. In contrast to WGA, the presence of a fucose residue α-1,6-linked to the GlcNAc residue involved in the N-glycosylamine bond does not affect the binding. "Bisected" hybrid-type N-glycosylpeptides are bound very tightly on immobilized rice lectin; but neither "bisected" nor "nonbisected" N-acetyllactosamine-type triantennary glycopeptides, substituted with GlcNAc residues at C-2 and C-4 of the α-1,3-linked mannose, interact with the immobilized lectin. In addition, two O-glycosidically linked oligosaccharides, containing one or two terminal nonreducing GlcNAc residues, weakly interact and are eluted as retarded fractions from the rice lectin column.

However, in this study, the interaction between the immobilized lectin and poly(N-acetyllactosamine)-type glycans was not investigated.

Summarizing, it appears that immobilized rice lectin presents a similar carbohydrate-binding specificity as immobilized WGA. However, at equal coupling density (5 mg/ml of gel), the rice lectin interacts with higher affinity with different N-glycosidically linked glycans, particularly with "bisected" biantennary N-glycosylpeptides, and represents a good alternative to immobilized WGA for the fractionation of this class of N-glycosylpeptides.

3.3.3 Subfractionation of N-Glycosylpeptides Using Immobilized Lectins with a Specificity Directed Toward Terminal Dominant-Oligosaccharidic Sequences

3.3.3.1 Subfractionation of Poly(N-Acetyllactosamine)-Type N-Glycosylpeptides and Related Oligosaccharides

Immobilized wheat germ agglutinin not only represents a very useful tool for the fractionation and analysis of poly(N-acetyllactosamine)-type N-glycosylpeptides (see Immobilized Wheat Germ Agglutinin), but also several other immobilized lectins interact with this class of glycans.

Immobilized Pokeweed Mitogens (PMW)

Five isolectins, Pa-1 to Pa-5, can be isolated from the roots of *Phytolacca americana* (Waxdal 1974; Yokoyama et al. 1976). Isolectin Pa-1 is a potent hemagglutinin and is mitogenic for both B and T murine lymphocytes. Isolectins Pa-2, 3, 4, and 5 are only T-cell mitogens.

The carbohydrate-binding specificity of pokeweed mitogens Pa-1 and Pa-2 was first studied by inhibition of hemagglutination and by quantitative inhibition induced by various oligosaccharides, glycopeptides, or glycoproteins (Yokoyama et al. 1978). Chitin oligosaccharides as well as band 3 glycoprotein of human erythrocyte and related N-acetylpeptides were good inhibitors, suggesting that the lectins recognized the core N,N'-diacetylchitobiose in N-glycosidically linked glycans and poly(N-acetyllactosamine)-type glycans.

Later, Irimura and Nicolson (1983) determined the ability of immobilized Pa-1, Pa-2 and Pa-4 isolectins to interact with different N-glycosylpeptides. The three isolectins, at a coupling density of 200 μg/ml of gel, possess similar binding specificities toward poly(N-acetyllactosamine)-type glycopeptides from human band 3 glycoprotein which are retained on the immobilized lectins. In contrast, N-acetyllactosamine-type bi- and triantennary N-glycosylpeptides isolated from porcine thyroglobulin do not interact with the lectins. From the observed lower binding of pokeweed mitogen isolectins to fetal erythrocytes, Irimura and Nicolson (1983) suggested that the isolectins preferentially bind to a highly branched blood group I-antigenic determinant that includes a GlcNAc β-1,6-Gal sequence. However, Katagiri et al. (1983) also using immobilized Pa-2 but at a higher coupling density (4 mg/ml of gel) showed that some oligomannosidic-type N-glycosylpeptides from porcine thyroglobulin with the Man α-1,2-Man α-1,2-Man β-1,3-Man β-1,4-GlcNAc β-1,4-Glc NAc β-1,N-Asn sequence were strongly retarded, whereas N-acetyllactosamine-type or hybrid-type N-glycosylpeptides had much weaker interaction with the immobilized lectin. They also showed that the core N,N'-diacetylchitobiose sequence was essential for the interaction with oligomannosidic-type glycopeptides, since treatment of these glycopeptides with endo-β-N-acetylglucosaminidase H cleaving the N,N'-diacetylchitobiosyl moiety releases glycans which no longer interact with the immobilized Pa-2 lectin.

Immobilized Datura stramonium Agglutinin (DSA)

A lectin with both hemagglutinating and mitogenic activity has been purified from the seeds of the Solanaceae *Datura stramonium* (thorn apple or jimsonweed) (for a review, see Goldstein and Poretz 1986). Inhibition by various oligosaccharides of DSA-induced hemagglutination had shown that N,N'-diacetylchitobiose or N,N',N''-triacetylchitotriose were poor inhibitors of DSA (Debray et al. 1981), but that N,N',N'',N'''-tetra-acetylchitotetraose was five fold more inhibitory than chitotriose (Kilpatrick and Yeoman 1978), suggesting an extended carbohydrate-binding site for the lectin. Later, Crowley et al. (1984) reported that N-acetyllactosamine Gal β-1,4-GlcNAc was a good inhibitor of the quantitative precipitation of asialofetuin induced by DSA; The pentasaccharide Gal β-1,4-GlcNAc β-1,6 (Gal β-1,4-GlcNAcβ-1,2). Man was 3-fold more inhibitory than N-acetyllactosamine and 25-fold more inhibitory than the isomeric pentasaccharide Galβ-1,4-GlcNAcβ-1,4 (Galβ1,4-GlcNAcβ-1,2) Man.

Using immobilized *Datura stramonium* agglutinin, Cummings and Kornfeld (1984) confirmed that N-acetyllactosamine tri- and tetraantennary N-glycosylpeptides containing an outer mannose residue, substituted at the C-2 and C-6 positions by N-acetyllactosamine bound with high affinity and were eluted with a N,N'-diacetylchitobiose/N,N',N''-triacetylchitotriose solution (8 mg/ml). In contrast, biantennary glycopeptides or triantennary glycopeptides containing an outer mannose residue substituted at the C-2 and C-4 positions by N-acetyllactosamine do not bind to immobilized DSA. Desialylation of the interacting glycopeptides does not affect binding to the immobilized lectin, but galactose residues are required for high affinity interaction.

From this study, which confirmed the results of Crowley et al. (1984), Cummings and Kornfeld (1984) suggested that the *Datura* lectin presents the same specificity as L₄-PHA. However, they showed that immobilized DSA can also interact with high affinity with N-glycosylpeptides containing the linear, unbranched, repeating (Galβ-1,4-GlcNAcβ-1,3-Gal) poly(N-acetyllactosamine)-type sequence (blood group i antigenic determinant) and represents a very useful tool for the fractionation and study of glycans containing this sequence. The interaction of immobilized DSA with the poly(N-acetyllactosamine)-type, highly branched blood group I antigenic determinant with the GlcNAcβ-1,6-Gal sequence, was not investigated in their study.

In a later study, Yamashita et al. (1987) extended these results and investigated the behavior of reduced N-acetyllactosamine-type glycans on immobilized DSA. Their results indicated that the interactions of oligosaccharides with immobilized DSA are more complex than reported by Cummings and Kornfeld (1984). N-acetyllactosamine-type oligosaccharides can be separated into three fractions. Biantennary oligosaccharides do not interact with the lectin and can be separated from triantennary glycans with the pentasaccharidic sequence Galβ-1,4-GlcNAc β1,4 (Galβ-1,4-GlcNAcβ-1,2) Man, which are retarded and eluted with the equilibration buffer. Tri- and tetraantennary oligosaccharides with the pentasaccharidic sequence Galβ-1,4-GlcNAcβ-1,6 (Galβ1,4-GlcNAcβ-1,2) Man as well as bi-, tri- and tetraantennary oligosaccharides with at least one repeating Galβ-1,4-GlcNAc sequence on an outer antenna are strongly bound and are eluted with a 1 % solution of chitotriose and chitotetraose in the equilibration buffer. α-2,6/3 sialylation of the galactose residue(s) and α-1,3 fucosylation of the N-acetylglucosamine residue(s) decrease or inhibit the interaction of the oligosaccharides with immobilized DSA.

Interaction with the immobilized lectin is not affected by the inner core moiety of N-acetyllactosamine-type oligosaccharides or by the presence of a bisecting N-acetylglucosamine residue. Summarizing, this last study clearly shows that after calibration with standard oligosaccharides, an immobilized DSA column represents another useful tool for the fractionation and analysis of N-acetyllactosamine-type glycans with or without repeating N-acetyllactosamine sequences.

Immobilized Lyopersicon esculentum Agglutinin (Tomato Lectin)

According to hemagglutination inhibition studies, another Solanaceae lectin isolated from tomato juice and flesh was shown to also present an extended carbohydrate-binding site complementary to N, N', N'', N'''-tetraacetylchitotetraose (Nachbar et al. 1980). More recently, Merkle and Cummings (1987b) showed that immobilized tomato lectin interacted with high affinity with linear poly (N-acetyllactosamine)-type N-glycosylpeptides containing three or more Galβ-1,4-GlcNAcβ-1,3 repeating sequences. However, neither the effects of substitution, by sialic acid or fucose residues, of galactose or N-acetylglucosamine residues belonging to these repeating sequences nor the behavior of branched poly (N-acetyllactosamine)-type glycans with blood group I antigenic determinant were investigated in this study. Therefore, further analyses of the structural requirements involved in the interactions of immobilized tomato lectin with N-glycosylpeptides and related oligosaccharides are necessary before using this lectin for fractionation of glycans.

3.3.3.2 Subfractionation of N-Acetyllactosamine-Type N-Glycosylpeptides and Related Oligosaccharides

Immobilized Ricinus communis Agglutinin I (RCA I)

Castor bean seeds *(Ricinus communis)* contain two distinct galactose-binding lectins, RCA I and RCA II. RCA I is a strong hemagglutinin, whereas RCA II is a strong toxin with weak hemagglutinating activity. Both lectins bind to the terminal N-acetyllactosamine sequence, but an important point is that RCA II is inhibited by N-acetylgalactosamine whereas RCA I is not (for review, see Goldstein and Poretz 1986).

The structural requirements for interaction of N-glycosylpeptides and related oligosaccharides with RCA I were first studied with a precipitation assay and by Scatchard plot analysis of the obtained saturation curves (Baenziger and Fiete 1979b), or by inhibition of RCA I-induced agglutination of human erythrocytes (Debray et al. 1981). These studies showed that RCA I recognized terminal nonreducing β-1,4-linked galactose residues and to a lesser extent, β-1,3-linked galactose residues. The strength of the interaction increases with the number of terminal β-galactose residues. Substitution of these galactose by α-2,6-linked NeuAc residues decreases the affinity of RCA I for the glycans and their substitution by α-2,3-linked NeuAc residues inhibits the interaction. This phenomenon may be related to the higher rotational freedom of the α-2,6-linkage, leaving the N-acetyllactosamine sequences more accessible to the lectin (see Sect. 3.1.2.3). The presence of an α-1,6-linked fucose residue at the innermost GlcNAc residue of the core does not interfere with the binding of RCA I.

These results were confirmed and extended by further studies of the interaction of N-glycosylpeptides and related oligosaccharides with immobilized RCA I (Kornfeld et al. 1981, Irimura et al. 1981, Debray et al. 1983, Narasimhan et al. 1986, Harada et al. 1987, Green et al. 1987a). These studies show that the affinity between immobilized RCA I and N-acetyllactosamine-type glycans increases with the number of terminal β-galactose residues. Depending on the coupling density of lectin, mono-, di-, tri- and tetragalactosylated glycans can be differentially retarded, or bound and eluted with a 0.1 M lactose solution. Substitution of the N-acetyllactosamine sequences by NeuAc residues at the C-6 position of galactose decreases the affinity of RCA I, but substitution by NeuAc residues at the C-3 position of galactose inhibits glycan-RCA I interactions. Substitution of the core β-mannose by a "bisecting" GlcNAc residue does not interfere with the binding to the immobilized lectin (Green et al. 1987a) or it slightly enhances this interaction between N-glycosylpeptides and immobilized RCA I (Narasimhan et al. 1986). However, the immobilized lectin presents a higher affinity for "bisected" glycopeptides with a single β-galactose residue on the Galβ-1,4-GlcNAcβ-1,2-Manα-1,6 antenna than for the isomer containing the galactose residue on the α-1,3-Man-linked antenna (Narasimhan et al. 1986, Harada et al. 1987). In contrast, "nonbisected" monogalactosylated glycans with a galactose residue on the α-1,6-Man-linked antenna interact slightly less with RCA I than the glycans with the galactose residue on the α-1,3-Man-linked antenna (Harada et al. 1987, Green et al. 1987a).

Thus, as previously pointed out for immobilized L₄- and E₄-PHA or DSA, application of immobilized RCA I in the fractionation and analysis of N-acetyllactosamine-type N-glycosylpeptides or related glycans requires a careful calibration of the lectin column to be used with glycan standards. As many interacting glycans can only be more or less retarded, the use of long columns of the immobilized lectin is also recommended.

Immobilized Ricinus communis Agglutinin II (RCA II)

RCA II agglutinin interacts with β-galactose residues of N-acetyllactosamine-type glycans, but also with terminal N-acetylgalactosamine residues (Baenziger and Fiete 1979b, Debray et al. 1981).

More recently, the structural determinants required for interaction of glycans with RCA II were more precisely determined by lectin affinity high performance liquid chromatography (Green et al. 1987a).

As immobilized RCA I, RCA II strongly interacts with N-acetyllactosamine-type oligosaccharides possessing terminal β-1,4-linked galactose residues. However, in contrast to RCA I, bound oligosaccharides can only be eluted with a 0.2 M GalNAc solution in 1 % acetic acid. The presence of either α-2,6 or α-2,3-linked NeuAc residues equally decreases the interaction with RCA II.

In contrast to immobilized RCA I, RCA II can also strongly interact with N-acetyllactosamine-type oligosaccharides containting terminal, nonreducing β-1,4-linked GalNAc residues. However, sulfatation of these β-1,4-linked GalNAc residues decreases the interaction with RCA II.

Whereas RCA I presents weak affinity for N-acetyllactosamine-type oligosaccharides with one or two terminal β-1,3-linked galactose residues, immobilized RCA II strongly binds there glycans, requiring elution with a GalNAc solution. Immobilized RCA II interacts with higher affinity with an N-acetyllactosamine sequence belonging to the α-1,6-Man-linked antenna in contrast to RCA I which preferentially recognizes the N-acetyllactosamine sequence belonging to the α-1,3-Man-linked antenna.

In contrast with RCA I, the presence of an antenna linked to the C-6 position of the α-1,6-linked core mannose, found in tri'- and tetraantennary N-acetyllactosamine-type glycans decreases the interaction with immobilized RCA II. Last, but not least, RCA II can also interact with agalactobi- or triantennary oligosaccharides as long as there is no β-1,6-linked GlcNAc residue attached to the α-1,6-linked core mannose. All these results obtained by Green et al. (1987a), clearly demonstrate that immobilized RCA I and RCA II present distinct oligosaccharide specificities and that the combination of these two lectins allows the fractionation and analysis of N-acetyllactosamine-type glycans according to the nature and the linkage of their terminal sugar moieties as well as to the number of their outer antennae.

Immobilized Erythrina Agglutinins

Agglutinins have been purified from the seeds of different *Erythrina* species, belonging to the family Leguminosae (for a review, see Goldstein and Poretz 1986).

The specificity of *Erythrina cristagalli* agglutinin as well as of the lectins isolated from several other *Erythrina* species were examined by inhibition of agglutination, by a precipitation test using blood group substances with I and i activities or by spectroscopic methods (Bhattacharyya et al. 1981, Iglesias et al. 1982, Kaladas et al. 1982, De Boeck et al. 1984, Lis et al. 1985). The lectins were demonstrated to be specific for the unsubstituted Gal β-1,4-GlcNAc sequence present in the non sialylated antennae of N-acetyllactosamine-type glycans. Recently, the behavior of N-acetyllactosamine-type N-glycosylpeptides and related oligosaccharides on columns of four different *Erythrina* agglutinins (i. e., *E. cristagalli, E. corallodendron, E. latissima* and *E. lysistemon*) which were immobilized on Sepharose, was investigated (Debray et al. 1986). With the same coupling density (3 mg lectin/ml gel), the carbohydrate-binding specificities of the four lectins are very similar and are directed toward unmasked N-acetyllactosamine sequences. The main difference between the four lectins is their relative strength of interaction with a given glycan.

Generally, the affinity of the lectins increases with the number of N-acetyllactosamine sequences accessible to the lectins. For example, asialo-N-acetyllactosamine-type bi-, tri-, and tetraantennary oligosaccharides can be separated on an immobilized *Erythrina lysistemon* column. Weakly interacting glycans are differentially eluted and retarded with the equilibration buffer and strongly bound glycans are eluted with a 0.15 M galactose solution. Substitution of the N-acetyllactosamine sequences by sialic acid residues, either at the C-3 or C-6 position of galactose completely abolishes the affinity of the immobilized lectins for the oligosaccharides.

It is noteworthy that immobilized *Erythrina lysistemon* interacts with higher affinity with a tri'-antennary N-acetyllactosamine-type oligosaccharide possessing an N-acetyllactosamine sequence attached to the C-6 position of the α-1,6-linked core mannose than with the isomeric oligosaccharide possessing a N-acetyllactosamine sequence attached to the C-4 position of the α-1,3-linked core mannose. In this respect the *Erythrina lysistemon* lectin is similar to RCA I (see Immobilized *Ricinus communis* Agglutinin [RCAI]).

The presence of one or several Fucα-1,3-GlcNAc groups on outer branches decreases or completely abolishes the interaction between N-glycosylpeptides and the immobilized *Erythrina* lectins. This influence of Fucα-1,3-GlcNAc groups on binding to Con A, E_4- and L_4-PHA has been previously reported (Yamashita et al. 1980, Santer et al. 1983).

Substitution of the core β-mannose residue by a "bisecting" N-acetylglucosamine residue decreases the affinity of the *Erythrina* lectins. In this respect, the *Erythrina* lectins are also similar to immobilized RCA I and Con A (Debray et al. 1983; Narasimhan et al. 1986). Immobilized *Erythrina* lectins possess a weak affinity toward some hybrid-type glycopeptides with a Galβ-1,4-GlcNAc sequence accessible to the lectins.

These results show that, after careful calibration with well defined oligosaccharides and glycopeptides, the immobilized *Erythrina* agglutinins also represent valuable tools for the fractionation of N-acetyllactosamine-containing oligosaccharides. However, the interaction of immobilized *Erythrina* lectins with N-acetyllactosamine-type glycans depends on the coupling density and improvement in the separation of weakly reacting, retarded fractions can be obtained using thin and long lectin columns.

Immobilized Allomyrina dichotoma Lectins (Allo-A)

Two β-galactose-binding isolectins Allo-A I and Allo-A II have been isolated from the hemolymph of the beetle *Allomyrina dichotoma* (Umetsu et al. 1984). The immobilized lectin was shown to interact specifically with some glycoproteins from human serum (Umetsu et al. 1985). More recently, the carbohydrate-binding specificity of the lectin was investigated by analyzing the behavior of different glycopeptides and oligosaccharides on either immobilized unseparated isolectins (Sueyoshi et al. 1988b) or on immobilized Allo-A II isolectin (Yamashita et al. 1988). Both studies show that sialylated N-acetyllactosamine-type N-glycosylpeptides and related oligosaccharides bind to the immobilized lectin and are eluted with a lactose solution. Asialoglycans are only retarded on the lectin column and asialogalactoglycans do not interact with the lectin. These results show that the presence of Galβ-1,4-GlcNAc sequences is essential in the interaction of a glycan with the lectin and that the affinity of the lectin increases with the number of terminal Galβ-1,4-GlcNAc sequences. A biantennary oligosaccharide containing two Galβ-1,3-GlcNAc sequences does not interact with immobilized Allo-A II showing the specificity of the lectin for N-acetyllactosamine residues. The non-binding of sialyllactitol suggests that an intact terminal reducing glucose or N-acetylglucosamine residue in this sequence is necessary for recognition by the *Allomyrina* lectins.

The presence of a "bisecting" GlcNAc residue decreases the affinity of the immobilized lectins for N-acetyllactosamine-type oligosaccharides. Either linear or branched poly(N-acetyllactosamine)-type oligosaccharides with repeating 3 Galβ-1,4-GlcNAc β-1 sequences strongly interact with the immobilized lectins. As previously described for immobilized RCA I (see Immobilized *Ricinus communis* Agglutinin I) immobilized Allo A lectins preferentially recognize glycans containing a core α-mannose residue substituted at the C-2 and C-6 positions with N-acetyllactosamine sequences rather than with glycans substituted with the same sequences, but at the C-2 and C-4 positions.

Substitution of the galactose residues of the Galβ-1,4-GlcNAc sequences by α-2,6-linked NeuAc residues enhances the affinity of immobilized Allo-A II for N-acetyllactosamine-type or milk oligosaccharides. In contrast, substitution of the galactose residues by α-2,3-linked NeuAc, α-1,3-linked galactose or α-1,2-linked fucose residues inhibits the interaction with the lectin. Glycans with Galβ-1,4-(Fuc α-1,3) GlcNAc groups in their outer chain moieties do not interact with the lectin. Either N-acetyllactosamine-type N-glycosylpeptides or related glycans, with a reducing or a reduced GlcNAc residue, released by hydrazinolysis give the same elution profiles. However, some reduced N-acetyllactosamine-type oligosaccharides, isolated from urines of patients with lysosomal diseases, possessing the Man β-1,4-GlcNAc-ol sequence at their reducing termini, are more strongly retarded.

In contrast to Sueyoshi et al. (1988b), Yamashita et al. (1988) showed that an α-1,6-linked fucose residue at the innermost GlcNAc residue does not interfere in the interaction of N-acetyllactosamine-type glycopeptides or glycans with the immobilized lectins.

Moreover, Sueyoshi et al. (1988b) showed that immobilized Allo-A can also interact with sialylated or neutral hybrid-type N-glycosylpeptides possessing a Galβ-1,4-GlcNAc sequence, but that it has no affinity for the Galβ-1,3-GalNAc core sequence found in mucin-type glycoproteins.

In summary, immobilized *Allomyrina dichotoma* lectins represent not only a new, promising tool for the fractionation of glycopeptides or oligosaccharides containing at least one unmasked N-acetyllactosamine sequence, but also for the separation of N-acetyllactosamine-type glycans with a NeuAc α-2,6-Gal β-1,4-GlcNAc terminal sequence from their isomeric forms containing the NeuAc α-2,3-Gal β-1,4-GlcNAc sequence.

3.3.3.3 Subfractionation of Sialylated N-Acetyllactosamine-Type N-Glycosylpeptides and Related Oligosaccharides

Sialic acid-binding lectins appear to be mainly present in invertebrates, including the lectins of *Limulus polyphemus* (horseshoe crab), *Carcinoscorpius rotunda cauda* (Indian horseshoe crab), and *Limax flavus* (slug) (for a review, see Goldstein and Poretz 1986). However, although some of these lectins have been used to fractionate glycoproteins on the basis of their sialic acid content (Mohan et al. 1981), they were not reported to be efficient tools in the fractionation of sialylated glycopeptides or oligosaccharides, according to the nature of the sialic acid residues or to their sialic acid content.

However, two lectins isolated from elderberry bark (*Sambucus nigra* agglutinin, SNA) by Broekaert et al. (1984) and from *Maackia amurensis* seeds (MAL) by Kawaguchi et al. (1974) were recently found to specifically bind sialylated glycopeptides and oligosaccharides.

Immobilized Sambucus nigra Agglutinin (SNA)

The lactose-specific lectin from elderberry bark was first isolated in 1984 by Broekaert et al. Competitive sugar inhibition experiments showed that both D-galactose and N-acetyl-D-galactosamine were weak inhibitors of SNA. Neither N-acetylneuraminic acid nor N-glycolylneuraminic acid were inhibitors (Shibuya et al. 1987 a).
The carbohydrate-binding properties of immobilized SNA were then determined using standard glycans (Shibuya et al. 1987 b).
Oligosaccharides, glycopeptides or glycoproteins containing NeuAc α-2,6-Gal/Gal-NAc sequences bind to immobilized SNA and are eluted with 50–100 mM lactose, whereas those with NeuAc α-2,3-Gal/GalNAc do not interact with the lectin. Asialo derivatives of these compounds also pass through the column unretained. Furthermore, immobilized SNA can fractionate glycans according to the number of NeuAc α-2,6-Gal units they contain.

Immobilized Maackia amurensis Leukoagglutinin (MAL)

Seeds from the leguminous plant *Maackia amurensis* were found to contain two hemagglutinins (Kawaguchi et al. 1974). One of them weakly agglutinates human erythrocytes and is strongly mitogenic while the other one is a strong hemagglutinin. The strongly mitogenic *M. amurensis* hemagglutinin (MAM, according to Kawaguchi et al. 1974) was found by Wang and Cummings (1987) to be a potent leukoagglutinin (MAL) for the mouse lymphoma cell line BW 5147. Leukoagglutination was inhibited by low concentrations of 2,3-sialyllactose, but was not inhibited by either 2-6, sialyllactose, lactose, or free NeuAc.
The carbohydrate-binding properties of immobilized *Maackia amurensis* leuko-agglutinin (MAL) were then determined using standard glycopeptides (Wang and Cummings 1988). This study showed that immobilized MAL interacts with high affinity with N-acetyllactosamine-type glycopeptides containing the terminal NeuAc α-2,3-Gal β-1,4-GlcNAc sequence. These glycopeptides are strongly retarded on the lectin column. Asialoglycopeptides pass unretained through the column as well as N-glycosylpeptides with terminal sialic acid residues, but in α-2,6-linkage to galactose.
Therefore, immobilized *Maackia amurensis* leukoagglutinin (MAL) and *Sambucus nigra* agglutinin (SNA) represent two promising lectins which could be sequentially used to fractionate sialylated glycans on the basis of their sialic acid linkages and on the number of sialic acid residues they contain.

3.3.4 Subfractionation of N-Glycosylpeptides Using Immobilized Lectins with a Specificity Directed Toward Terminal Nonreducing Dominant Monosaccharides

3.3.4.1 Immobilized Griffonia simplicifolia Isolectins I (GSA I)

The seeds of *Griffonia simplicifolia* contain five tetrameric isolectins composed of two subunits, A and B (Murphy and Goldstein 1977). The carbohydrate-binding specificity of the two subunits differs significantly; the A subunit exhibits a primary specificity for α-linked N-acetyl-D-galactosamine residues, but also reacts with α-linked D-galactose residues: the B subunit presents a sharp specificity toward α-D-galactose residues. Immobilized GSA I isolectins were used to isolate either α-D-galactose-containing glycopeptides from blood group B erythrocytes or α-N-acetyl-D-galactosamine-containing glycopeptides from blood group A erythrocytes (Finne et al. 1978). A column of immobilized GSA I isolectins was also used to separate UDP-glucose, non retained and eluted at the void volume, from UDP-galactose which was strongly retarded (Blake and Goldstein 1980). The same column can also separate the corresponding 2-acetamido-2-deoxy derivatives, the alditols of lactose which is non-retained and of melibiose (Gal α-1,6-Glc), which is retarded on the column. In the same study, Blake and Goldstein also showed that immobilized GSA I-B$_4$, with a sharp specificity for terminal α-D-galactose residues, can separate UDP-N-acetyl-D-galactosamine, which is very weakly retarded on the column from strongly retarded UDP-galactose. Affinity chromatography of N-acetyllactosamine-type glycopeptides from bovine thyroglobulin on immobilized GSA I isolectins allows the separation of glycopeptides, containing a single terminal, nonreducing α-D-galactose residue which are is weakly retarded on the lectin column in contrast to glycopeptides containing multiple α-D-galactose residues which are firmly bound and eluted with a 10 mM methyl α-D-galactose solution (Spiro and Bhoyroo 1984). GSA I isolectins also bind glycans containing terminal nonreducing N-acetyl-D-galactosamine residues.

Recently, GSA I isolectins, noncovalently bound through their oligomannosidic-type glycans to covalently immobilized concanavalin A (Con A-Sepharose), were used to fractionate glycolipid-derived oligosaccharides from rabbit erythrocytes according to their number of terminal α-D-galactose residues (Wang et al. 1988).

3.3.4.2 Immobilized Griffonia simplicifolia Agglutinin II (GSA II)

Another lectin isolated from the seeds of *Griffonia simplicifolia* by affinity chromatography on chitin was shown to be a N-acetyl-D-glucosamine-specific agglutinin (Iyer et al. 1976). According to quantitative precipitation studies, GSA II was shown to present the stronger affinity for glycoconjugates with terminal, nonreducing α- or β-linked N-acetyl-D-glucosamine residues (Ebisu et al. 1978).

The immobilized GSA II represents a very useful tool for the fractionation of glycopeptides or oligosaccharides with terminal, nonreducing GlcNAc residues (Debray et al. 1983). The affinity between immobilized GSA II and oligosaccharides increases with the number of terminal, nonreducing GlcNAc residues. With a low coupling density of lectin (2 mg of lectin per ml of gel), at least one GlcNAc residue is necessary to obtain a weak interaction between glycans and the immobilized lectin, giving a retarded elution. The elution volume increases with the number of terminal, nonreducing GlcNAc residues and the oligosaccharides with four terminal GlcNAc residues are strongly bound and eluted with a 0.15 M N-acetyl-D-glucosamine solution (Debray et al. 1983).

3.3.4.3 Immobilized Galanthus nivalis (Snowdrop) Agglutinin (GNA)

Recently, a lectin was isolated from snowdrop *(Galanthus nivalis)* bulbs by affinity chromatography on immobilized mannose. This lectin presents an exclusive specificity toward mannose (Van Damme et al. 1987). The carbohydrate-binding properties of this new lectin were examined by inhibition of precipitation of *Hansenula capsulata* mannan with oligosaccharides of known structures and by affinity chromatography of different glycoproteins, polysaccharides and glycoasparagines on the immobilized lectin (Shibuya et al. 1988 b).

Their results show that GNA presents a strict requirement for terminal, nonreducing mannose residues and a preference for Man α-1,3-Man terminal groups. The immobilized GNA interacts only with some glycoasparagines carrying this terminal, nonreducing Man α-1,3-Man sequence, giving a retarded elution. The interaction with the lectin also depends on the number of these disaccharide units. However, hybrid-type glycopeptides possessing this Man α-1,3-Man sequence together with a "bisecting" GlcNAc residue do not interact with the immobilized lectin, probably due to steric hindrance.

It is noteworthy that another mannose-specific lectin, isolated from tulip bulbs *(Tulipa gesneriana)* by Oda and Minami (1986) was found to be more effectively inhibited by Man α-1,6-linked oligosaccharides. However, a detailed carbohydrate-binding study of this lectin, by affinity chromatography of standard oligosaccharides on the immobilized lectin, has not yet been performed.

3.3.4.4 Immobilized Bowringia milbraedii Agglutinin (BMA)

Another mannose-specific lectin was recently purified from the seeds of the Nigerian legume *Bowringia milbraedii* (Animashaun and Hughes 1989). The carbohydrate-binding specificity of the lectin was studied by affinity chromatography of known glycans on the immobilized lectin. Immobilized BMA presents a higher affinity for oligosaccharides with a terminal, nonreducing Man α-1,2-Man sequence. Moreover high affinity binding requires that at least a Man α-1,2-Man α-1,6-Man α-1,6-Man sequence, present in a typical Man_9 $GlcNAc_2$ oligomannosidic-type glycan (see Fig. 3.2). This glycan as well as the $Man_8GlcNAc$ and $Man_7GlcNAc$ oligosaccharides possessing the tetrasaccharide sequence are tightly bound and their elution requires a 0.2 M methyl α-D-mannoside solution. In contrast, other $Man_8GlcNAc$ and $Man_7GlcNAc$ isomers as well as $Man_6GlcNAc$ and $Man_5GlcNAc$, in which this sequence is lacking, are more weakly bound and are eluted with a 0.01 M methyl α-D-mannoside solution. However, the behavior of the corresponding oligomannosidic-type glycopeptides on immobilized BMA was not analyzed in this study.

In summary, immobilized *Galanthus nivalis* and *Bowringia milbraeddi* agglutinins represent, in addition to Concanavalin A, promising lectins which could be sequentially used for the fractionation of oligomannosidic-type glycans according to the linkage of their terminal, nonreducing mannose residues.

3.3.4.5 Affinity Chromatography on Immobilized Fucose-Binding Lectins

Fucose-binding lectins have been isolated from several sources, including *Ulex europaeus, Lotus tetragonolobus, Evonymus europaeus* and *Griffonia simplicifolia* seeds as well as from eel serum *(Anguilla anguilla)* and from the *Aleuria aurantia* mushroom (for a review, see Goldstein and Poretz 1986). The specificity of these lectins toward L-fucose-containing oligosaccharides has been studied either by inhibition of hemagglutination or by inhibition of quantitative precipitation. However, only the ability of immobilized *Ulex europaeus* (UEAI), *Lotus tetragonolobus* (LTA) and *Aleuria aurantia* (AAA) agglutinins to interact with L-fucose-containing glycans was investigated.

Immobilized LTA and UEAI present no affinity for fucosylated N-acetyllactosamine-type N-glycosylpeptides or related oligosaccharides. Only the glycopeptide Fuc α-1,6-GlcNAc β-1,N-Asn or the fucodisaccharide Fuc α-1,6-GlcNAc were found to be either retarded on immobilized LTA (Susz and Dawson 1979) or bound and eluted with a 0.1 M L-fucose solution (Montreuil et al. 1986).

However, another fucose-specific lectin, isolated from the mushroom *Aleuria aurantia* (Kochibe and Furukawa 1980), was demonstrated to represent a very effective tool for the fractionation of glycopeptides and oligosaccharides according to their fucose content (Yamashita et al. 1985; Amano et al. 1985; Harada et al. 1987; Debray and Montreuil 1989).

These studies show that immobilized *Aleuria aurantia* agglutinin strongly interacts with all N-glycosylpeptides or related glycans containing a fucose residue in an α-1,6-linkage to the innermost GlcNAc residue of the core. It is noteworthy that glycans, released by hydrazinolysis are still recognized by the immobilized lectin. From this point of view, immobilized AAA differs from immobilized LCA, PSA or VFA, in which such glycans, released by hydrazinolysis and still possessing the α-1,6-linked L-fucosyl determinant, no longer interact with the immobilized lectins (Debray and Montreuil 1983; Katagiri et al. 1984).

Immobilized AAA weakly interacts with glycans containing a fucose residue in the α-1,3-linkage to one of the outer GlcNAc residues in the N-acetyllactosamine-type glycans. The presence of both an α-1,6- and an α-1,3-linked fucose enhances the affinity of the lectin for the glycans, whereas the interaction is slightly decreased by the presence of a "bisecting" GlcNAc residue (Debray and Montreuil 1989). In addition, oligosaccharides, isolated from human milk and from the urine of patients with fucosidosis and containing either the Fuc α-1,2-Gal β-1,4-GlcNAc or Gal β-1,4 (Fuc α-1,3) GlcNAc sequences, interact with the immobilized lectin but to a lesser extent than N-acetyllactosamine-type glycans possessing an α-1,6-linked fucose residue in the core. Oligosaccharides with a Gal β-1,3 (Fuc α-1,4) GlcNAc sequence interact less strongly than oligosaccharides containing the above two groups. Finally, oligosaccharides with the Fuc α-1,2-Gal β-1,3-GlcNAc or Gal β-1,4-GlcNAc β-1,3-Gal β-1,4 (Fuc α-1,3) GlcNAc sequences do not interact with the immobilized *Aleuria* lectin (Yamashita et al. 1985).

In conclusion, after careful calibration with well defined oligosaccharides and glycopeptides, the immobilized *Aleuria aurantia* agglutinin presently represents the only valuable tool for the fractionation and the resolution of the microheterogeneity of glycans due to the presence of different L-fucosyl substituents.

A approach in the use of immobilized Con A, LCA and other lectins to sequentially fractionate N-glycosylpeptides and glycans was described previously (Montreuil et al. 1986).

3.3.5 Development of High Performance Affinity Chromatography of Nucleotide Sugars and of Glycopeptides and Glycans on Immobilized Lectins

The use of the lectin affinity high-performance liquid chromatography (HPLC) technique for the separation of nucleotide sugars on columns of poly(acrylic ester)gel (WG 003)-bound lectins was proposed by Tokuda et al. (1985). For example, a *Ricinus communis* agglutinin I (RCA I) WG 003 column was found to represent a very effective affinity adsorbent for the rapid separation of UDP-Gal or UDP-N-acetyl-D-galactosamine which interacted with the immobilized lectin, from other nucleotide-sugars such as UDP-N-acetyl-D-glucosamine. More recently, Green et al. (1987 a) have shown that lectin affinity HPLC, using columns of silica-bound lectins, was particularly useful in two types of study. The first one is for defining the oligosaccharide specificity of lectins. For example, Green et al. (1987b) found that oligosaccharide specificities displayed by silica-bound leukoagglutinating phytohemagglutinin (L_4-PHA), Con A and *Datura stramonium* agglutinin (DSA) were almost identical to those established with the same lectins immobilized on agarose. However, further studies by Baenziger's group have shown that subtle changes in glycan structures could be detected by lectin affinity HPLC (Green et al. 1987b, 1988; Green and Baenziger 1987). As pointed out by these authors lectin affinity HPLC could also be used to purify glycopeptides or glycans by using preparative columns. For example, asialooligosaccharides released from human IgG have been separated into five fractions with a *Ricinus communis* agglutinin I WG 003 column described above (Harada et al. 1987). The operation time was reported to be approximately one third of that by the conventional RCA I-agarose column. As proposed for affinity chromatography using lectins immobilized on agarose, most of the N-glycosylpeptides or related glycans could be fractionated into homogeneous classes by serial affinity HPLC using different columns of well-defined, silica-bound lectins.

3.3.6 Fractionation of Mucin-Type Glycosylpeptides or O-Glycosidically-Linked Oligosaccharides with Immobilized Lectins

Although lectin affinity chromatography is now an extensively used method for the fractionation of N-glycosylpeptides or related glycans, very few applications can be found in the literature on the separation of mucin-type glycopeptides or O-glycosidically linked oligosaccharides. Only immobilized wheat germ agglutinin was found to interact specifically with glycopeptides containing a high density of O-glycosidically linked sialyloligosaccharides (Bhavanandan et al. 1977; Furukawa et al. 1986). A disialyltetrasaccharide-alditol (compound 4 of Fig. 3.1) was found to weakly interact with a WGA-Sepharose column with a high WGA content (Furukawa et al. 1986). However, by inhibition of hemagglutination or by inhibition of quantitative precipitation, many lectins were shown to bind to mucin-type oligosaccharides. These lectins have been reviewed in a recent classification according to their specificities toward N-acetyl-D-galactosamine or galactose (Wu et al. 1988).

Recently, the carbohydrate-binding specificities of five of these lectins: ABA-I isolated from *Agaricus bisporus* (Presant and Kornfeld 1972; Sueyoshi et al. 1985), PNA isolated from *Arachis hypogaea* (Lotan and Sharon 1978), BPA isolated from *Bauhinia purpurea* (Osawa et al. 1978), SBA isolated from *Glycine max* (Lis and Sharon 1972), and VVA-B$_4$ isolated from *Vicia villosa* seeds (Tollefsen and Kornfeld 1983 a, b) have been studied by affinity chromatography on the immobilized lectins of mucin-type glycopeptides and related oligosaccharides of known structures (Sueyoshi et al. 1988 a). The association constants of the five lectins for these structures were also determined in this study by frontal affinity chromatography according to the method of Kasai and Ishii (1978 a, b).

The five lectins can be classified into two groups on the basis of their reactivity with either Gal β-1,3-GalNAc α-1,3-Ser/Thr recognized by ABA-I, PNA or BPA or GalNAc α-1,3-Ser/Thr recognized by SBA and VVA-B$_4$ with a higher affinity than the Gal β-1,3-GalNAc α-1,3-Ser/Thr sequence.

It is also noteworthy that only the immobilized *Agaricus bisporus* agglutinin binds a sialylated glycopeptide from human erythrocyte glycophorin A containing three tetrasaccharide chains with the NeuAc α-2,3-Gal β-1,3 [NeuAc α-2,6-] GalNAc α-1,3 sequence. Immobilized soybean agglutinin (SBA) was shown to preferentially interact with oligosaccharides containing the GalNAc α-1,3-Gal β-1,3-GlcNAc sequence, isolated from a blood group A-active substance.

However, the affinity of immobilized ABA, PNA or BPA for the disaccharide Gal β-1,3-GalNAc-ol is quite low. This represents a limiting factor in the use of the three immobilized lectins and especially of an ABA-Sepharose column for the fractionation of sialylated oligosaccharide-alditols released from mucin-type glycoproteins by reductive cleavage of the O-glycosidic linkages.

The search for new lectins capable of recognizing determinants on more complex mucin-type glycopeptides or related oligosaccharide-alditols must be continued in order to complete the powerful tool represented by serial lectin affinity chromatography in the fractionation of N-glycosylpeptides and related glycans.

3.4 Affinity Chromatography of Glycolipids on Immobilized Lectins

Until recently, the use of lectins in the study of glycolipids was limited to the interaction of ^{125}I-labeled lectins with glycosphingolipids, separated by thin layer chromatography, by the overlay technique developed by Magnani et al. (1980), for detection of gangliosides that bind cholera toxin. This method has allowed the determination of glycolipid specificity of several lectins including peanut agglutinin (Momoi et al. 1982), *Helix pomatia* lectin (Smith 1983), *Vicia villosa* B 4 isolectin (Bailly et al. 1985), *Erythrina cristagalli* and soybean agglutinin (Ehrlich-Rogozinski et al. 1987), and wheat germ agglutinin (Higashi et al. 1988).

Another solid-phase method derived from ELISA (enzyme-linked immunosorbent assay) was also developed to determine glycosphingolipid-lectin interaction (Molin et al. 1986; Ehrlich-Rogozinski et al. 1987). It involves the binding of biotin-conjugated lectins to glycolipids immobilized on the plastic surfaces of microtiter plates in the presence of lecithin and cholesterol and quantification of the bound lectin by incubation with an avidin-horseradish peroxidase solution. However, in contrast to the overlay method, this technique can only be used to follow the interaction of individual purified glycosphingolipids with a lectin.

Another approach is to follow the interaction between liposomes containing individual purified glycosphingolipids and the lectin immobilized either on Sepharose or Affigel 10 (Boldt et al. 1977; Maget-Dana et al. 1981; Månsson and Olofsson 1983).

As described in the above references, these different methods have allowed to clarify the specificities of different lectins toward lipid-bound carbohydrate structures. However, the most promising method for purification of glycolipids by affinity chromatography on immobilized lectins was recently proposed by Torres and Smith (1988). Their method relies on the retention of the carbohydrate-binding specificity of an immobilized lectin in aqueous solutions of tetrahydrofuran, which disrupts glycolipid micelle formation. Glycolipids are bound to the immobilized lectin in a solvent containing 95 % tetrahydrofuran and 5 % water; after application of a step gradient increasing the water content up to 50 %, the specifically bound glycolipids are eluted in solvent containing the competive monosaccharide.

The lectin affinity system described by Torres and Smith (1988) allowed them to purify glycolipids containing terminal, nonreducing α-1,3-GalNAc residues, such as the Forssman and human blood group A-active glycolipids, on immobilized *Helix pomatia* lectin.

As pointed out by the authors, this method will represent a powerful tool either for the purification of glycolipids or for the determination of lectin specificity toward different glycosphingolipids, but on the condition that the carbohydrate-binding activity and specificity of lectins should be retained in the mobile phase selected to destroy glycolipid micelle formation.

3.5   References

Allen AK, Neuberger A, Sharon N (1973) The purification, composition and specificity of wheat-germ agglutinin. Biochem J 131 : 155–162

Amano J, Messer M, Kobata A (1985) Structures of the oligosaccharides isolated from milk of the platypus. Glycoconjugate J 2 : 121–135

Animashaun T, Hughes RC (1989) *Bowringia milbraedii* agglutinin: specificity of binding to early processing intermediates of asparagine-linked oligosaccharide and use as a marker of endoplasmic reticulum glycoproteins. J Biol Chem 264 : 4657–4663

Aono S, Sato H, Semba R, Kashiwamata S (1985) Improved separation at low temperature of glycoproteins by Con A-Sepharose affinity chromatography in the presence of sodium dodecyl sulfate (SDS). Experientia 41 : 1084–1087

Aucouturier P, Mihaesco E, Mihaesco C, Preud'Homme JL (1987) Characterization of Jacalin, the human IgA and IgD binding lectin from jackfruit. Mol Immunol 24 : 503–511

Aucouturier P, Duarte F, Mihaesco E, Pineau N, Preud'Homme JL (1988) Jacalin, the human IgA$_1$ and IgD precipitating lectin, also binds IgA$_2$ of both allotypes. J Immunol Methods 113 : 185–191

Azevedo Moreira R, Ainouz IL (1981) Lectins from the seeds of jackfruit *(Artocarpus integrifolia L)*: isolation and purification of two isolectins from the albumin fraction. Biol Plant 23 : 186–192

Baenziger JU, Fiete D (1979 a) Structural determinants of Concanavalin A specificity for oligosaccharides. J Biol Chem 254 : 2400–2407

Baenziger JU, Fiete D (1979b) Structural determinants of *Ricinus communis* agglutinin and toxin specificity for oligosaccharides. J Biol Chem 254:9795–9799

Baenziger JU, Kornfeld S (1974) Structure of the carbohydrate units of the IgA$_1$ immunoglobulin II. Structure of the O-glycosidically linked oligosaccharide units. J Biol Chem 249:7270–7281

Bailly P, Tollefsen SE, Cartron JP (1985) Glycolipid specificity of the B$_4$ lectin from *Vicia villosa* seeds. Glycoconjugate J 2:401–408

Baumstark JS (1983) Guidelines for the preparative fractionation of human serum proteins on gradient-eluted columns of Concanavalin A-Sepharose: elution positions of fourteen well-characterized proteins and evidence for Concanavalin A-reactive albumin IgA and IgG complexes. Prep Biochem 13:315–345

Bayard B, Kerckaert JP (1981) Uniformity of carbohydrate chains within molecular variants of rat α$_1$-fetoprotein with distinct affinity for Concanavalin A. Eur J Biochem 113:405–414

Bayard B, Kerckaert JP, Laine A, Hayem A (1982) Uniformity of glycans within molecular variants of α$_1$-protease inhibitor with distinct affinity for Concanavalin A. Eur J Biochem 124:371–376

Bessler W, Goldstein IJ (1973) Phytohemagglutinin purification: a general method involving affinity and gel chromatography. FEBS lett 34:58–61

Bhattacharyya L, Das PK, Sen A (1981) Purification and properties of D-galactose-binding lectins from some *Erythrina species*: comparison of properties of lectins from *E. indica, E. arborescens, E. suberosa* and *E. lithosperma*. Arch Biochem Biophys 211:459–470

Bhavanandan VP, Katlic AW (1979) The interaction of Wheat germ agglutinin with sialoglycoproteins. The role of sialic acid. J Biol Chem 254:4000–4008

Bhavanandan VP, Umemoto J, Banks JR, Davidson EA (1977) Isolation and partial characterization of sialoglycopeptides produced by a murine melanoma. Biochemistry 16:4426–4437

Bierhuizen MFA, Edzes HT, Schiphorst WECM, Van Den Eijnden DH, Van Dijk W (1988) Effect of α (2-6)-linked sialic acid and α (1-3)-linked fucose on the interaction of N-linked glycopeptides and related oligosaccharides with immobilized *Phaseolus vulgaris* leukoagglutinating lectin (L-PHA). Glycoconjugate J 5:85–97

Bittiger H, Schnebli HP (eds) (1976) Concanavalin A as a tool, Wiley. Lond, 639 pp

Blake DA, Goldstein IJ (1980) Resolution of nucleotide sugars and oligosaccharides by lectin affinity chromatography. Anal Biochem 102:103–109

Bøg-Hansen TC (1973) Crossed immuno-affinoelectrophoresis. An analytical method to predict the result of affinity chromatography. Anal Biochem 56:480–488

Bøg-Hansen TC (1983) Affinity electrophoresis of glycoproteins. In: Scouten WH (ed) Solid phase biochemistry: analytical synthetic aspects, Wiley, New York, pp 223–251

Bøg-Hansen TC, Bjerrum OJ, Brogren CH (1977) Identification and quantification of glycoproteins by affinity electrophoresis. Anal Biochem 81:78–87

Boldt DH, Speckart SF, Richards RL, Alving CR (1977) Interactions of plant lectins with glycolipids in liposomes. Biochem Biophys Res Commun 74:208–214

Broekaert WF, Nsimba-Lubaki M, Peeters B, Peumans WJ (1984) A lectin from elder (*Sambucus nigra* L.) bark. Biochem J 221:163–169

Carter WG, Sharon N (1977) Properties of the human erythrocyte membrane receptors for peanut and *Dolichos biflorus* lectins. Arch Biochem Biophys 180:570–582

Carver JP, Brisson JR (1984) The three-dimensional structure of N-linked oligosaccharides. In: Ginsburg V, Robbins PW (eds) Biology of carbohydrates, Wiley, New York, Vol 2, pp 289–331

Clemetson KJ, Pfueller SL, Luscher EF, Jenkins CSP (1977) Isolation of the membrane glycoproteins of human blood platelets by lectin affinity chromatography. Biochim Biophys Acta 464:493–508

Crowley JF, Goldstein IJ, Arnarp J, Lönngren J (1984) Carbohydrate binding studies on the lectin from *Datura stramonium* seeds. Arch Biochem Biophys 231:524–533

Cummings RD, Kornfeld S (1982a) Characterization of the structural determinants required for the high affinity interaction of asparagine-linked oligosaccharides with immobilized *Phaseolus vulgaris* leuko-agglutinating and erythroagglutinating lectins. J Biol Chem 257:11230–11234

Cummings RD, Kornfeld S (1982b) Fractionation of asparagine-linked oligosaccharides by serial lectin-Agarose affinity chromatography. A rapid, sensitive and specific technique. J Biol Chem 257:11235–11240

Cummings RD, Kornfeld S (1984) The distribution of repeating [Gal β 1,4 GlcNAc β 1,3] sequences in asparagine-linked oligosaccharides of the mouse lymphoma cell lines BW 5147 and PHA[R] 2.1. Binding of oligosaccharides containing these sequences to immobilized *Datura stramonium* agglutinin. J Biol Chem 259:6253–6260

De Boeck H, Loontiens FG, Lis H, Sharon N (1984) Binding of simple carbohydrates and some N-acetyllac-tosamine-containing oligosaccharides to *Erythrina cristagalli* agglutinin as followed with a fluorescent indicator ligand. Arch Biochem Biophys 234:297–304

Debray H, Montreuil J (1981) Structural basis for the affinity of four insolubilized lectins, with a specificity for α-D-mannose, towards various glycopeptides with the N-glycosylamine linkage and related oligosaccharides. In: Bøg-Hansen TC (ed) Lectins: biology, biochemistry, clinical biochemistry, De Gruyter, Berlin (W), Vol 1, pp 221–230

Debray H, Montreuil J (1983) Structural basis for the affinity of four insolubilized lectins, with a specificity for α-D-mannose, towards various glycopeptides with the N-glycosylamine linkage and related oligosaccharides. J Biosci 5 (suppl.) 1:93–100

Debray H, Montreuil J (1989) *Aleuria aurantia* agglutinin. A new isolation procedure and further study of its specificity towards various glycopeptides and oligosaccharides. Carbohydr Res 185:15–26

Debray H, Decout D, Strecker G, Spik G, Montreuil J (1981) Specificity of twelve lectins towards oligosaccharides and glycopeptides related to N-glycosylproteins. Eur J Biochem 117:41–55

Debray H, Pierce-Crétel A, Spik G, Montreuil J (1983) Affinity of ten insolubilized lectins towards various glycopeptides with the N-glycosylamine linkage and related oligosaccharides. In: Bøg-Hansen TC, Spengler GA (eds) Lectins: biology, biochemistry, clinical biochemistry, De Gruyter, Berlin (W), Vol 3, pp 335–350

Debray H, Montreuil J, Lis H, Sharon N (1986) Affinity of four immobilized *Erythrina* lectins toward various N-linked glycopeptides and related oligosaccharides. Carbohydr Res 151:359–370

Dorai DT, Bachhawat BK, Bishayee S (1981) Fractionation of sialoglycoproteins on an immobilized sialic acid-binding lectin. Anal Biochem 115:130–137

Dulaney JT (1979) Binding interactions of glycoproteins with lectins. Mol cell Biochem 21:43–63

Ebisu S, Iyer PN, Goldstein IJ (1978) Equilibrium dialysis and carbohydrate-binding studies on the 2-acetamido-2-deoxy-D-glucopyranosyl-binding lectin from *Bandeiraea simplicifolia* seeds. Carbohydr Res 61:129–138

Ehrlich-Rogozinski S, De Maio A, Lis H, Sharon N (1987) The glycolipid specificity of *Erythrina cristagalli* agglutinin. Glycoconjugate J 4:379–390

Ernst-Cabrera K, Wilchek M (1987) From affinity chromatography to HPAC. In: Burgess R (ed), Protein purifications: micro to macro, Liss, New York, pp 163–175

Finne J, Krusius T (1982) Preparation and fractionation of glycopeptides. Methods Enzymol 83:269–277

Finne J, Krusius T, Rauvala H, Kekomäki R, Myllylä G (1978) Alkali-stable blood group A- and B-active poly(glycosyl)-peptides from human erythrocyte membrane. FEBS Lett 89:111–115

Finne J, Krusius T, Jarnefelt J (1980) Fractionation of glycopeptides. In: Varmavuori A (ed) 27th Int Congr of Pure and Applied Chemistry, Pergamon Press, Oxford New York, pp 147–159

Furukawa K, Minor JE, Hegarty JD, Bhavanandan VP (1986) Interaction of sialoglycoproteins with wheat germ agglutinin-Sepharose of varying ratio of lectin to Sepharose. Use for the purification of mucin glycoproteins from membrane extracts. J Biol Chem 261:7755–7761

Gallagher JT (1984) Carbohydrate-binding properties of lectins: a possible approach to lectin nomenclature and classification. Biosci Rep 4:621–632

Gallagher JT, Morris A, Dexter TM (1985) Identification of two binding sites for wheat-germ agglutinin on polylactosamine-type oligosaccharides. Biochem J 231:115–122

Gleeney JR, Walborg EF (1979) Lectin affinity chromatography of cell surface proteins of Novikoff tumor cells. J Supramol Struct 11:493–502

Goldstein IJ, Hayes CE (1978) The lectins: carbohydrate-binding proteins of plants and animals. Adv Carbohydrate Chem Biochem 35:127–340

Goldstein IJ, Poretz RD (1986) Isolation, physicochemical characterization and carbohydrate-binding specificity of lectins: In: Liener IE, Sharon N, Goldstein IJ (eds), The lectins: properties, functions and applications in biology and medicine, Academic Press, London, New York pp 33–247

Goldstein IJ, Hammarström S, Sundblad S (1975) Precipitation and carbohydrate-binding specificity studies on wheat germ agglutinin. Biochim Biophys Acta 405:53–61

Goldstein IJ, Hughes RC, Monsigny M, Osawa T, Sharon N (1980) What should be called a lectin? Nature (Lond) 285:66

Green ED, Baenziger JU (1987) Oligosaccharide specificities of *Phaseolus vulgaris* leukoagglutinating and erythroagglutinating phytohemagglutinins. Interactions with N-glycanase-released oligosaccharides. J Biol Chem 262:12 018–12 029

Green ED, Brodbeck RM, Baenziger JU (1987 a) Lectin affinity high-performance liquid chromatography. Interactions of N-glycanase-released oligosaccharides with *Ricinus communis* agglutinin I and *Ricinus communis* agglutinin II. J Biol Chem 262:12 030–12 039

Green ED, Brodbeck RM, Baenziger JU (1987 b) Lectin affinity high-performance liquid chromatography: interactions of N-glycanase-released oligosaccharides with leukoagglutinating phytohemagglutinin, Concanavalin A, *Datura stramonium* agglutinin and *Vicia villosa* agglutinin. Anal Biochem 167:62–75

Green ED, Adelt G, Baenziger JU, Wilson S, Van Halbeek H (1988) The asparagine-linked oligosaccharides on bovine fetuine. Structural analysis of N-glycanase-released oligosaccharides by 500-megahertz ^{1}H NMR spectroscopy. J Biol Chem 263:18 253–18 268

Hammarström S, Hammarström ML, Sundblad G, Arnarp J, Lönngren J (1982) Mitogenic leukoagglutinin from *Phaseolus vulgaris* binds to a pentasaccharide unit in N-acetyllactosamine-type glycoprotein glycans. Proc Natl Acad Sci USA 79:1611–1615

Harada H, Kamei M, Tokumoto Y, Yui S, Koyama F, Kochibe N, Endo T, Kobata A (1987) Systematic fractionation of oligosaccharides of human immunoglobulin G by serial affinity chromatography on immobilized lectin columns. Anal Biochem 164:374–381

Harboe M, Saltvedt E, Closs O, Olsnes S (1975) Interactions between *Ricinus* agglutinin and human plasma proteins. Scand J Immunol 4 (suppl) 2:125–134

Hayman MJ, Crumpton MJ (1972) Isolation of glycoproteins from pig lymphocyte plasma membrane using *Lens culinaris* phytohemagglutinin. Biochem Biophys Res Commun 47:923–930

Hayman MJ, Skehel JJ, Crumpton MJ (1973) Purification of virus glycoproteins by affinity chromatography using *Lens culinaris* phytohemagglutinin. FEBS Lett 29:185–188

Hedo JA (1984) Lectins as tools for the purification of membrane receptors. In: Venter C, Harrison L (eds) Receptor purification procedures, Liss, New York, Vol 2, pp 45–60

Hedo JA, Harrison LC, Roth J (1981) Binding of insulin receptors to lectins: evidence for common carbohydrate determinants on several membrane receptors. Biochemistry 20:3385–3393

Higashi H, Sugii T, Kato S (1988) Specific staining on thin-layer chromatograms of glycosphingolipids of neolactoseries and gangliosides with a terminal N-acetylneuraminyl residue by different procedures with wheat germ agglutinin. Biochim Biophys Acta 963:333–339

Iglesias JL, Lis H, Sharon N (1982) Purification and properties of a D-galactose/N-acetyl-D-galactosamine-specific lectin from *Erythrina cristagalli*. Eur J Biochem 123:247–252

Imam A, Laurence DJR, Neville AM (1981) Isolation and characterization of a major glycoprotein from milk-fat globule membrane of human breast milk. Biochem J 193:47–54

Irimura T, Nicolson GL (1983) Interaction of pokeweed mitogen with poly (N-acetyllactosamine)-type carbohydrate chains. Carbohydr Res 120:187–195

Irimura T, Tsuji T, Tagami S, Yamamoto K, Osawa T (1981) Structure of a complex-type sugar chain of human glycophorin A. Biochemistry 20:560–566

Ivatt RJ, Harnett PB, Reeder JW (1986a) Isolated erythroglycans have a high-affinity interaction with wheat germ agglutinin but are poorly accessible in situ. Biochim Biophys Acta 881:124–134

Ivatt RJ, Reeder JW, Clark GF (1986b) Structural and conformational features that affect the interaction of polylactosamino-glycans with immobilized wheat germ agglutinin. Biochim Biophys Acta 883:253–264

Iwase H, Hotta K (1977) Ovotransferrin subfractionation dependent upon carbohydrate chain differences. J Biol Chem 252:5437–5443

Iwase H, Kato Y, Hotta K (1981) Ovalbumin subfractionation and individual difference in ovalbumin microheterogeneity. J Biol Chem 256:5638–5642

Iyer PNS, Wilkinson KD, Goldstein IJ (1976) An N-acetyl-D-glucosamine binding lectin from *Bandeiraea simplicifolia* seeds. Arch Biochem Biophys 177:330–333

Junqua S, Lemonnier M, Robert L (1981) Glycoconjugates from *"Spongia officinalis" (Phylum porifera)*. Isolation, fractionation by affinity chromatography on lectins and partial characterization. Comp Biochem Physiol 69B:445–453

Kahane I, Furthmayr H, Marchesi VT (1976) Isolation of membrane glycoproteins by affinity chromatography in the presence of detergents. Biochim Biophys Acta 426:464–476

Kaladas PM, Kabat EA, Iglesias JL, Lis H, Sharon N (1982) Immunochemical studies on the combining site of the D-galactose/N-acetyl-D-galactosamine specific lectin from *Erythrina cristagalli* seeds. Arch Biochem Biophys 217:624–637

Kasai K, Ishii S (1978a) Affinity chromatography of trypsin and related enzymes. V. Basic studies of quantitative affinity chromatography. J Biochem (Tokyo) 84:1051–1060

Kasai K, Ishii S (1978b) Studies on the interaction of immobilized trypsin and specific ligands by quantitative affinity chromatography. J Biochem (Tokyo) 84:1061–1069

Katagiri Y, Yamamoto K, Tsuji T, Osawa T (1983) Structural requirements for the binding of high-mannose-type glycopeptides to immobilized pokeweed Pa-2 lectin. Carbohydr Res 120:283–292

Katagiri Y, Yamamoto K, Tsuji T, Osawa T (1984) Structural requirements for the binding of glycopeptides to immobilized *Vicia faba* (fava) lectin. Carbohydr Res 129:257–265

Kato Y, Iwase H, Hotta K (1984) Preparation of eight ovalbumin subfractions by combined lectin affinity chromatography. Anal Biochem 138:437–441

Kàwaguchi T, Matsumoto I, Osawa T (1974) Studies on hemagglutinins from *Maackia amurensis* seeds. J Biol Chem 249:2786–2792

Kennedy JF, Rosevear A (1973) An assessment of the fractionation of carbohydrates on Concanavalin A – Sepharose 4B by affinity chromatography. J Chem Soc Perkin Trans pp 2041–2046

Kerckaert JP, Bayard B (1982) Glycan uniformity within molecular variants of transferrin with distinct affinity for Concanavalin A. Biochem Biophys Res Commun 105:1023–1030

Kilpatrick DC, Yeoman MM (1978) Purification of the lectin from *Datura stramonium*. Biochem J 175:1151–1153

Kobata A (1984) The carbohydrates of glycoproteins. In: Ginsburg V, Robbins PW (eds) Biology of carbohydrates, Wiley, New York, Vol 2, pp 87–161

Kobayashi K, Kondoh H, Hagiwara K, Vaerman JP (1988) Jacalin: chaos in its immunoglobulin-binding specificity. Mol Immunol 25:1037–1038

Kochibe N, Furukawa K (1980) Purification and properties of a novel fucose-specific hemagglutinin of *Aleuria aurantia*. Biochemistry 19:2841–2846

Kohn J, Wilchek M (1982) A new approach (cyano-transfer) for cyanogen bromide activation of Sepharose at neutral pH, which yields activate resins, free of interfering nitrogen derivatives. Biochem Biophys Res Commun 107:878–884

Koide N, Muramatsu T (1974) Endo-β-N-acetylglucosaminidase acting on carbohydrate moieties of glycoproteins. Purification and properties of the enzyme from *Diplococcus pneumonia*. J Biol Chem 249:4897–4904

Kondoh H, Kobayashi K, Hagiwara K (1987) A simple procedure for the isolation of human secretory IgA of IgA₁ and IgA₂ subclass by a jackfruit lectin, Jacalin, affinity chromatography. Mol Immunol 24:1219–1222

Kornfeld K, Reitman ML, Kornfeld R (1981) The carbohydrate-binding specificity of pea and lentil lectins. Fucose is an important determinant. J Biol Chem 256:6633–6640

Kornfeld R, Kornfeld S (1970) The structure of a phytohemagglutinin receptor site from human erythrocytes. J Biol Chem 245:2536–2545

Kornfeld R, Ferris C (1975) Interactions of immunoglobulin glycopeptides with Concanavalin A. J Biol Chem 250:2614–2619

Kornfeld R, Kornfeld S (1985) Assembly of asparagine-linked oligosaccharides. Ann Rev Biochem 54:631–664

Kornfeld S, Kornfeld R (1978) Use of lectins in the study of mammalian glycoproteins. In: Horowitz M, Pigman W (eds) The glycoconjugates, Academic Press, Lond New York, Vol 2, pp 437–449

Kronis KA, Carver JP (1982) Specificity of isolectins of wheat germ agglutinin for sialyloligosaccharides: a 360 – MHz proton nuclear magnetic resonance binding study. Biochemistry 21:3050–3057

Krusius T, Finne J, Rauvala H (1976) The structural basis of the different affinities of two types of acidic N-glycosidic glycopeptides for Concanavalin A-Sepharose. FEBS Lett 71:117–120

Larsson PO, Glad M, Hansson L, Mansson MO, Ohlson S, Mosbach K (1983) High-performance liquid affinity chromatography. Adv Chromatogr 21:41–85

Lis H, Sharon N (1972) Soy bean *(Glycine max)* agglutinin. Methods Enzymol 28:360–368

Lis H, Sharon N (1977) Lectins: their chemistry and applications to immunology. In: Sela M (ed) The antigens, Academic Press, Lond New York, Vol 4, pp 429–529

Lis H, Sharon N (1981) Lectins in higher plants. In: Marcus A (ed) The biochemistry of plants, Academic press, New York, Vol 6, pp 371–447

Lis H, Sharon N (1984) Lectins: properties and applications to the study of complex carbohydrates in solution and on cell surfaces. In: Ginsburg V, Robbins PW (eds) Biology of carbohydrates, Wiley, New York, Vol 2, pp 2–85

Lis H, Sharon N (1986 a) Applications of lectins. In: Liener IE, Sharon N, Goldstein IJ (eds) The lectins, properties, functions and applications in biology and medicine. Academic Press, Lond New York, pp 265–291

Lis H, Sharon N (1986 b) Lectins as molecules and as tools. Ann Rev Biochem 55:35–67

Lis H, Joubert FJ, Sharon N (1985) Isolation and properties of N-acetyllactosamine specific lectins from nine *Erythrina* species. Phytochemistry 24:2803–2809

Lotan R, Nicolson GL (1979) Purification of cell membrane glycoproteins by lectin affinity chromatography. Biochim Biophys Acta 559:329–376

Lotan R, Sharon N (1978) Peanut *(Arachis hypogaea)* agglutinin. Methods Enzymol 50:361–367

Lotan R, Beattie G, Hubbell W, Nicolson GL (1977) Activities of lectins and their immobilized derivatives in detergent solutions. Implications on the use of lectin affinity chromatography for the purification of membrane glycoproteins. Biochemistry 16:1787–1794

Maget-Dana R, Veh RW, Sander M, Roche AC, Schauer R, Monsigny M (1981) Specificities of limulin and wheat-germ agglutinin towards some derivatives of GM₃ gangliosides. Eur J Biochem 114:11–16

Magnani JL, Smith DF, Ginsburg V (1980) Detection of gangliosides that bind cholera toxin: direct binding of ¹²⁵I-labeled toxin to thin-layer chromatograms. Anal Biochem 109:399–402

Manabe T, Higuchi N, Okuyama T (1988) High-performance affinity chromatography of human serum Concanavalin A binding proteins. J Chromatogr 431:45–54

Månsson JE, Olofsson S (1983) Binding specificities of the lectins from *Helix pomatia*, soybean and peanut against different glycosphingolipids in liposome membranes. FEBS Lett 156:249–252

March SC, Parikh I, Cuatrecasas P (1974) A simplified method for cyanogen bromide activation of Agarose for affinity chromatography. Anal Biochem 60:149–152

Matsuura S, Chen HC (1980) A simple and effective solvent system for elution of gonadotropins from Concanavalin A affinity chromatography. Anal Biochem 106:402–410

Mellis SJ, Baenziger JU (1983a) Structures of the oligosaccharides present at the three asparagine-linked glycosylation sites of human IgD. J Biol Chem 258:11 546–11 556

Mellis SJ, Baenziger JU (1983b) Structures of the O-glycosidically linked oligosaccharides of human IgD. J Biol Chem 258:11 557–11 563

Merkle RK, Cummings RD (1987a) Lectin affinity chromatography of glycopeptides. Methods Enzymol 138:232–259

Merkle RK, Cummings RD (1987b) Relationship of the terminal sequences to the length of poly-N-acetyllactosamine chains in asparagine-linked oligosaccharides from the mouse lymphoma cell line BW 5147. Immobilized tomato lectin interacts with high affinity with glycopeptides containing long poly-N-acetyllactosamine chains. J Biol Chem 262:8179–8189

Mohan S, Bishayee S, Bachhawat BK (1981) Resolution of microheterogeneity in rat liver acid phosphatase using immobilized sialic acid binding lectin. Indian J Biochem Biophys 18:177–181

Molin K, Fredman P, Svennerholm L (1986) Binding specificities of the lectins PNA, WGA and UEAI to polyvinylchloride-adsorbed glycosphingolipids. FEBS Lett 205:51–55

Momoi T, Tokunaga T, Nagai Y (1982) Specific interaction of peanut agglutinin with the glycolipid asialo GM1. FEBS Lett 141:6–10

Monsigny M, Delmotte F, Helene C (1978) Ligands containing heavy atoms: perturbation of phosphorescence of a tryptophan residue in the binding site of wheat germ agglutinin. Proc Natl Acad Sci USA 75:1324–1328

Monsigny M, Roche AC, Sené C, Maget-Dana R, Delmotte F (1980) Sugar-lectin interactions: how does wheat-germ agglutinin bind sialoglycoconjugates? Eur J Biochem 104:147–153

Montreuil J (1980) Primary structure of glycoprotein glycans. Basis for the molecular biology of glycoproteins. Adv Carbohydr Chem Biochem 37:157–223

Montreuil J (1982) Glycoproteins. In: Neuberger A, Van Deenen LLM (eds) Comprehensive biochemistry, Elsevier, Amsterdam, Vol 19B, Part II, pp 1–188

Montreuil J (1983) Conformation of the glycoprotein glycans of the N-acetyllactosaminic type (complex type). Biochem Soc Trans 11:134–136

Montreuil J (1984a) Spatial structures of glycan chains of glycoproteins in relation to metabolism and function. Survey of a decade of research. Pure Appl Chem 56:859–877

Montreuil J (1984b) Spatial conformation of glycans and glycoproteins Biol cell 51:115–132

Montreuil J, Fournet B, Spik G, Strecker G (1978) Etude theorique de la conformation spaciale des glycannes de la sérotransferrine humaine. C R Acad Sci Paris Ser D 287:837–840

Montreuil J, Debray H, Debeire P, Delannoy P (1983) Lectins as oligosaccharide receptors. In: Popper H, Reutter W, Köttgen E, Gudat F (eds) Structural carbohydrates in the liver, Falk Symp 34, MTP Press, Boston, pp 239–258

Montreuil J, Bouquelet S, Debray H, Fournet B, Spik G, Strecker G (1986) Glycoproteins. In: Chaplin MF, Kennedy JF (eds) Carbohydrate analysis. A practical approach, IRL Press, Oxford, pp 143–204

Murphy LA, Goldstein IJ (1977) Five α-D-galactopyranosyl-binding isolectins from Bandeiraea simplicifolia seeds. J Biol Chem 252:4739–4742

Nachbar MS, Oppenheim JD, Thomas JO (1980) Lectins in the U.S. diet. Isolation and characterization of a lectin from the tomato (Lycopersicon esculentum). J Biol Chem 255:2056–2061

Nakamura S, Tanaka K, Murkawa S (1960) Specific protein of legumes which reacts with animal proteins. Nature (Lond) 188:144–145

Narasimhan S, Wilson JR, Martin E, Schachter H (1979) A structural basis for four distinct elution profiles on Concanavalin A-Sepharose affinity chromatography of glycopeptides. Can J Biochem 57:83–96

Narasimhan S, Freed JC, Schachter H (1986) The effect of a "bisecting" N-acetylglucosaminyl group on the binding of biantennary, complex oligosaccharides to Concanavalin A, *Phaseolus vulgaris* erythroagglutinin (E-PHA), and *Ricinus communis* agglutinin (RCA-120) immobilized on Agarose. Carbohydr Res 149:65–83

Oda Y, Minami K (1986) Isolation and characterization of a lectin from tulip bulbs, *Tulipa gesneriana.* Eur J Biochem 159:239–245

Ogata S, Muramatsu T, Kobata A (1975) Fractionation of glycopeptides by affinity column chromatography on Concanavalin A-Sepharose. J Biochem (Tokyo) 78:687–696

Ohyama Y, Kasai K, Nomoto H, Inoue Y (1985) Frontal affinity chromatography of ovalbumin glycoasparagines on a Concanavalin A-Sepharose column. A quantitative study of the binding specificity of the lectin. J Biol Chem 260:6882–6887

Osawa T, Tsuji T (1987) Fractionation and structural assessment of oligosaccharides and glycopeptides by use of immobilized lectins. Ann Rev Biochem 56:21–42

Osawa T, Irimura T, Kawaguchi T (1978) *Bauhinia purpurea* agglutinin. Methods Enzymol 50:367–372

Pereira MEA, Kabat EA (1976) Immunochemical studies on blood groups. LXII. Fractionation of hog and human A, H and AH blood group active substances on insoluble immunoadsorbents of *Dolichos* and *Lotus* lectins. J Exp Med 143:422–436

Peters BP, Ebisu S, Goldstein IJ, Flashner M (1979) Interaction of wheat germ agglutinin with sialic acid. Biochemistry 18:5505–5511

Poliquin L, Shore DC (1980) A method for efficient and selective recovery of membrane glycoproteins from Concanavalin A-Sepharose using media containing sodium dodecyl sulfate and urea. Anal Biochem 109:460–465

Poola I, Narasimhan S (1988) Interaction of asparagine-linked oligosaccharides with an immobilized rice *(Oryza sativa)* lectin column. Biochem J 250:117–124

Poola I, Seshadri HS, Bhavanandan VP (1986) Purification and saccharide-binding characteristics of a rice lectin. Carbohydr Res 146:205–217

Presant CA, Kornfeld S (1972) Characterization of the cell surface receptor for the *Agaricus bisporus* hemagglutinin. J Biol Chem 247:6937–6945

Reinwald E, Rautenberg P, Risse HJ (1981) Purification of the variant antigens of *Trypanosoma congolese.* A new approach to the isolation of glycoproteins. Biochim Biophys Acta 668:119–131

Renkonen O, Mäkinen P, Hård K, Helin J, Penttilä L (1988) Immobilized wheat germ agglutinin separates small oligosaccharides derived from poly-N-acetyllactosaminoglycans of embryonal carcinoma cells. Biochem Cell Biol 66:449–453

Robey FA, Liu TY (1981) Limulin: AC-reactive protein from *Limulus polyphemus.* J Biol Chem 256:969–975

Roque-Barreira MC, Campos-Neto A (1985) Jacalin: an IgA-binding lectin. J Immunol 134:1740–1743

Saito M, Tohyoshima S, Osawa T (1978) Isolation and partial characterization of the major sialoglycoprotein of human T-lymphoblastoid cells of a MOLT-4B cell line. Biochem J 175:823–831

Samor B, Michalski JC, Debray H, Mazurier C, Goudemand M, Van Halbeek H, Vliegenthart JFG, Montreuil J (1986) Primary structure of a new tetraantennary glycan of the N-acetyllactosaminic type isolated from human factor VIII/Von Willebrand factor. Eur J Biochem 158:295–298

Santer UV, Glick M, Van Halbeck H, Vliegenthart JFG (1983) Characterization of the neutral glycopeptides containing the structure α-L-fucopyranosyl-(1→3)-2-acetamido-2-deoxy-D-glucose from human neuroblastoma cells. Carbohydr Res 120:197–213

Sastry MVK, Banarjee P, Patanjali SR, Swamy MJ, Swarnalatha GV, Surolia A (1986) Analysis of saccharide binding to *Artocarpus integrifolia* lectin reveals specific recognition of T-antigen (β-D-Gal (1→3) D-GalNAc). J Biol Chem 261:11 726–11 733

Scher MG, Resneck WG, Bloch RJ (1989) Stabilisation of immobilized lectin columns by crosslinking with glutaraldehyde. Anal Biochem 177:168–171

Shibata S, Peters BP, Roberts DD, Goldstein IJ, Liotta LA (1982) Isolation of laminin by affinity chromatography on immobilized *Griffonia simplicifolia* I lectin. FEBS Lett 142:194–198

Shibuya N, Goldstein IJ, Broekaert WF, Nsimba-Lubaki M, Peeters B, Peumans WJ (1987a) The elderberry (*Sambucus nigra* L.) bark lectin recognizes the NeuAc(α2-6)Gal/GalNAc sequence. J Biol Chem 262: 1596–1601

Shibuya N, Goldstein IJ, Broekaert WF, Nsimba-Lubaki M, Peeters B, Peumans WJ (1987b) Fractionation of sialylated oligosaccharides, glycopeptides and glycoproteins on immobilized elderberry (*Sambucus nigra* L.) bark lectin. Arch Biochem Biophys 254: 1–8

Shibuya N, Berry JE, Goldstein IJ (1988a) One-step purification of murine IgM and human α_2-macroglobulin by affinity chromatography on immobilized snowdrop bulb lectin. Arch Biochem Biophys 267: 676–680

Shibuya N, Goldstein IJ, Van Damme EJM, Peumans WJ (1988b) Binding properties of a mannose-specific lectin from the snowdrop (*Galanthus nivalis*) bulb. J Biol Chem 263: 728–734

Smith DF (1983) Glycolipid-lectin interactions: detection by direct binding of ^{125}I-lectins to thin layer chromatograms. Biochem Biophys Res Commun 115: 360–367

Spiro RG, Bhoyroo VD (1984) Occurrence of α-D-galactosyl residues in the thyroglobulins from several species. Localization in the saccharide chains of the complex carbohydrate units. J Biol Chem 259: 9858–9866

Strickler JE, Mancini PE, Patton CL (1978) *Trypanosoma brucei brucei*: isolation of the major surface coat glycoprotein by lectin affinity chromatography. Exp Pathol 46: 262–276

Sueyoshi S, Tsuji T, Osawa T (1985) Purification and characterization of four isolectins of mushroom *Agaricus bisporus*. Biol Chem Hoppe-Seyler 366: 213–221

Sueyoshi S, Tsuji T, Osawa T (1988a) Carbohydrate-binding specificities of five lectins that bind to O-glycosyl-linked carbohydrate chains. Quantitative analysis by frontal-affinity chromatography. Carbohydr Res 178: 213–224

Sueyoshi S, Yamamoto K, Osawa T (1988b) Carbohydrate binding specificity of a beetle (*Allomyrina dichotoma*) lectin. J Biochem (Tokyo) 103: 894–899

Susz JP, Dawson G (1979) The affinity of the fucose-binding lectin from *Lotus tetragonolobus* for glycopeptides and oligosaccharides accumulating in fucosidosis. J Neurochem 32: 1009–1013

Takasaki S, Kobata A (1974) Microdetermination of individual neutral and amino sugars and N-acetylneuraminic acid in complex saccharides. J Biochem (Tokyo) 76: 783–789

Tarago MT, Tucker KA, Van Halbeek H, Smith DF (1988) A novel sialylhexasaccharide from human milk: purification by affinity chromatography on immobilized wheat germ agglutinin. Arch Biochem Biophys 267: 353–362

Tokuda M, Kamei M, Yui S, Koyama F (1985) Rapid resolution of nucleotide sugars by lectin affinity high-performance liquid chromatography. J chromatogr 323: 434–438

Tollefsen SE, Kornfeld R (1983b) Isolation and characterization of lectins from *Vicia villosa*. Two distinct carbohydrate binding activities are present in seed extracts. J Biol Chem 258: 5165–5171

Tollefsen SE, Kornfeld R (1983a) The B$_4$ lectin from *Vicia villosa* seeds interacts with N-acetylgalactosamine residues α-linked to serine or threonine residues in cell surface glycoproteins. J Biol Chem 258: 5172–5176

Torres BV, Smith DF (1988) Purification of Forssman and human blood group A glycolipids by affinity chromatography on immobilized *Helix pomatia* lectin. Anal Biochem 170: 209–219

Umetsu K, Kosaka S, Suzuki T (1984) Purification and characterization of a lectin from the beetle *Allomyrina dichotoma*. J Biochem (Tokyo) 95: 239–245

Umetsu K, Ikeda N, Kashimura S, Suzuki T (1985) Affinity chromatography of human serum proteins using immobilized lectin from *Allomyrina dichotoma*. Biochem Int 10: 549–552

Van Damme EJM, Allen AK, Peumans WJ (1987) Isolation and characterization of a lectin with exclusive specificity towards mannose from snowdrop (*Galanthus nivalis*) bulbs. FEBS Lett 215: 140–144

Wang WC, Cummings RD (1987) An assay for leukoagglutinating lectins using suspension cultured mouse lymphoma cells (BW 5147) stained with neutral red. Anal Biochem 161: 80–84

Wang WC, Cummings RD (1988) The immobilized leukoagglutinin from the seeds of *Maackia amurensis*

95

binds with high affinity to complex-type Asn-linked oligosaccharides containing terminal sialic acid-linked α-2,3 to penultimate galactose residues. J Biol Chem 263:4576–4585

Wang WC, Clark GF, Smith DF, Cummings RD (1988) Separation of oligosaccharides containing terminal α-linked galactose residues by affinity chromatography on *Griffonia simplicifolia* I bound to Concanavalin A-Sepharose. Anal Biochem 175:390–396

Waxdal MJ (1974) Isolation, characterization and biological activities of five mitogens from pokeweed. Biochemistry 13:3671–3677

Wilchek M, Miron T, Kohn, J (1984) Affinity chromatography. Methods Enzymol 104:3–10

Wu AM (1984) Differential binding characteristics and applications of D-Gal β-1→3 D-GalNAc specific lectins. Mol Cell Biochem 61:131–141

Wu AM, Sugii S, Herp A (1988) A guide for carbohydrate specificities of lectins. Adv Exp Med Biol 228:819–847

Yamamoto K, Tsuji T, Matsumoto I, Osawa T (1981) Structural requirements for the binding of oligosaccharides and glycopeptides to immobilized wheat germ agglutinin. Biochemistry 20:5894–5899

Yamamoto K, Tsuji T, Osawa T (1981) Requirement of the core structure of a complex-type glycopeptide for the binding to immobilized lentil- and pea-lectins. Carbohydr Res 110:283–289

Yamashita K, Tachibana Y, Nakayama T, Kitamura M, Endo Y, Kobata A (1980) Structural studies of the sugar chains of human parotid α-amylase. J Biol Chem 255:5635–5642

Yamashita K, Hitoi A, Kobata A (1983) Structural determinants of *Phaseolus vulgaris* erythroagglutinating lectin for oligosaccharides. J Biol Chem 258:14 753–14 755

Yamashita K, Kochibe N, Ohkura T, Ueda I, Kobata A (1985) Fractionation of L-fucose-containing oligosaccharides on immobilized *Aleuria aurantia* lectin. J Biol Chem 260:4688–4693

Yamashita K, Totani K, Ohkura T, Takasaki S, Goldstein IJ, Kobata A (1987) Carbohydrate binding properties of complex-type oligosaccharides on immobilized *Datura stramonium* lectin. J Biol Chem 262:1602–1607

Yamashita K, Umetsu K, Suzuki T, Iwaki Y, Endo T, Kobata A (1988) Carbohydrate binding specificity of immobilized *Allomyrina dichotoma* lectin II. J Biol Chem 263:17 482–17 489

Yokoyama K, Yano O, Terao T, Osawa T (1976) Purification and biological activities of pokeweed *(Phytolacca americana)* mitogens. Biochim Biophys Acta 427:443–452

Yokoyama K, Terao T, Osawa T (1978) Carbohydrate-binding specificity of pokeweed mitogens. Biochim Biophys Acta 538:384–396

Zanetta JP, Reeber A, Vincendon G (1981) Glycoproteins from adult rat brain synaptic vesicles. Fractionation on four immobilized lectins. Biochim Biophys Acta 670:393–400

Zehr BD, Litwin SD (1987) Human IgD and IgA₁ compete for D-galactose-related binding sites on the lectin Jacalin. Scand J Immunol 26:229–236

4 The Kinetics of Lectin-Mediated Cell Agglutination

Christian Flemming

4.1 Introduction

The ability of most lectins to agglutinate suspended cells is a well known fact. However, the mechanism of cell agglutination is not fully understood at present. The overall reaction

cell + lectin + cell $\rightleftharpoons$ cell-lectin + cell $\rightleftharpoons$ cell-lectin-cell

only describes the formation of lectin bridges between suspended cells, which is a simplification of the complicated process of cell agglutination.

The study of agglutination kinetics can contribute to elucidation of the phenomenon of cell agglutination and to the quantification of the specific interactions between lectin molecules and cell surface receptors. In this report a model for lectin mediated cell agglutination is presented that is based on the following assumptions:

1. The theories for colloid flocculation can be applied to cell agglutination
2. The velocity of cell agglutination can be determined by the alteration of the optical density of the cell suspension
3. The lectin binding to the cell surface is described by a simple equilibrium process
4. The binding of lectin molecules to the cell surface is a fast reaction compared to the cell agglutination (equilibrium of lectin binding is accomplished before agglutination starts).

Though these preconditions are not exactly fulfilled the study of the agglutination kinetics is a useful tool for the characterization of lectin preparations, for comparison of different lectins with related specificities, and for determinations of the inhibition by competitive sugars.

On the other hand, different microorganism strains of the same species, mutants of one strain or mammalian cell lines can be characterized and differentiated by their specific agglutination properties. Moreover investigations of cell agglutination lead to a better understanding of the binding sites and binding forces, and the structures and dynamics of cell surfaces.

4.2 Theory

The following theoretical considerations are mainly based on the excellent works of Linnemans et al. (1976 a, b).

97

For large scattering particles in a diluted suspension the light extinction is given by

$$E = 0.434 \cdot d \cdot \Sigma n_j \cdot A_j \cdot K_j \tag{1}$$

Where E, denotes light extinction (optical density);

        d, cuvette path lenght;

        n, number of particles;

        A, cross-sectional area;

        K, scattering efficiency. Subscript j denotes the optical class.

During the process of cell agglutination the single cells associate forming agglutinates (clumps, flocs). The number of particles (n) in the system decreases from n_o to n_t (the cells in one agglutinate count for one particle) but the cross sectional area A of an agglutinate is larger than that of a single cell.

If we assume that cells and cell agglutinates are spherical and the cell population is homogeneous, the relation of the optical density at the start of the reaction and at the measuring point of time is given by

$$\frac{E_o}{E_t} = \frac{0.434 \cdot d \cdot n_o \cdot A_o \cdot K_j}{0.434 \cdot d \cdot n_t \cdot A_t \cdot K_j} = \frac{n_o \cdot A_o}{n_t \cdot A_t} \tag{2}$$

In the case of spheres the correlations between V, n and r are

$$V_o = n_o \cdot \frac{4}{3} \cdot \pi \cdot r_o^3 \quad \text{(single cells, at the start of the reaction)}$$

$$V_t = n_t \cdot \frac{4}{3} \cdot \pi \cdot r_t^3 \quad \text{(agglutinates)}$$

Where V denotes volume, r the radius. Subscript o denotes at the start of the reaction and t at the measuring time t.

It is postulated that the density of the cells does not change during the agglutination, i.e., the volume of all cells is constant ($V_o = V_t$).

$$n_o \cdot \frac{4}{3}\pi \cdot r_o^3 = V_o = V_t = n_t \cdot \frac{4}{3}\pi \cdot r_t^3 \text{ resp.}$$

$$\frac{r_o^3}{r_t^3} = \frac{n_t}{n_o}$$

The scattering area of a sphere is a circle with the same diameter. The value of this area is given by

$$A_o = \pi \cdot r_o^2 \quad \text{(for the single cell)}$$

$$A_t = \pi \cdot r_t^2 \quad \text{(for the agglutinate)}$$

This relation may be described by

$$\frac{A_o}{A_t} = \frac{r_o^2}{r_t^2} \tag{4}$$

By substitution of r_o and r_t in Eq. (3) we obtain the following expression

$$\frac{A_o}{A_t} = \frac{n_t^{2/3}}{n_o^{2/3}} \tag{5}$$

and by substitution of the values A_o and A_t in Eq. (2) the following equations can be given

$$\frac{E_o}{E_t} = \frac{n_o \cdot n_t^{2/3}}{n_t \cdot n_o^{2/3}} = \frac{n_o^{1/3}}{n_t^{1/3}} = \sqrt[3]{\frac{n_o}{n_t}} \text{ resp.}$$

$$\frac{n_o}{n_t} = \left(\frac{E_o}{E_t}\right)^3 = \tilde{n} \tag{6}$$

Where $\tilde{n}$ denotes the number of cells per agglutinate.

Thus means the average number of cells which are associated in one agglutinate, can be easily calculated from the optical density of the suspension. For example, if the light extinction decreases from 0.75 before agglutination to 0.375 during the agglutination we must write

$$\left(\frac{E_o}{E_t}\right)^3 = \left(\frac{0.75}{0.375}\right)^3 = 2^3 = 8$$

Theoretically one agglutinate is formed from eight single cells. The decrease in the optical density caused by cell agglutination is illustrated in Fig. 4.1 a.
In accordance to Linnemans et al. (1976 a, b) we describe the kinetics of lectin-mediated cell agglutination as a second-order process. With regard to the lectin con-

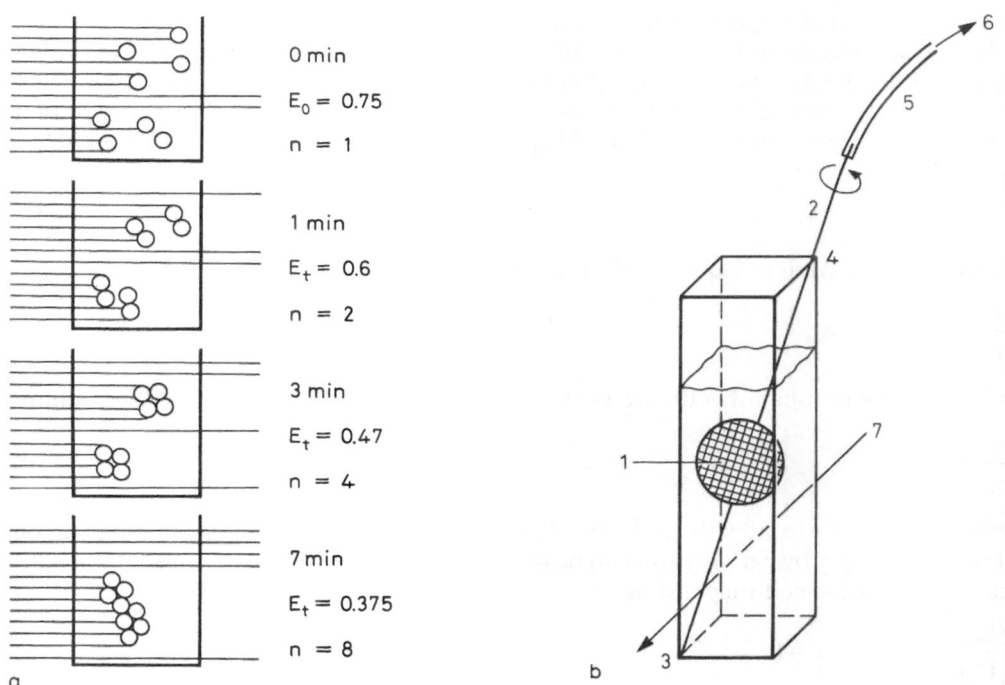

Fig. 4.1 Quantitative determination of cell agglutination
a) Principle of the method: single cells (circles) agglutinate, and more light beams (lines) pass the cuvette
b) Continously stirred cuvette: 1. disk of 8mm diameter; 2. steel needle; 3. bottom of cuvette; 4. top of cuvette; 5. small gum; 6. motor attachment; 7. light beam

centration it seems to be a saturation reaction. So the velocity of cell agglutination is given by

$$v = -\frac{d_n}{d_t} = k^\circ \cdot n^2 \cdot \theta \cdot (1 - \theta),$$

where v ist the true velocity of cell agglutination and the term θ expresses the fraction of occupied lectin binding sites, i. e., the relation between lectin-loaded receptors and the sum of all receptors. This value ranges from 0 to 1 and can be calculated with the Langmuir isotherm

$$\frac{\theta}{1 - \theta} = K_a \cdot C.$$

If we assume, that the first step of the reaction, i. e., the binding of the lectin, to be very fast and the second step, the agglutination to be quite slow, we can then conclude that θ is constant during the agglutination term. Thus

$$-\frac{dn}{dt} = k^+ \cdot n^2, \tag{8}$$

where k°, k^+ are constants; k is a constant of the velocity of cell agglutination $(\min^{-1})$; K_a is an equilibrium constant; C is an equilibrium concentration; and c ist the „apparent" velocity of cell agglutination $(\min^{-1})$ (see sect. 4.3 and fig. 4.2).

The cell agglutination in a stirred suspension progresses only to a certain extent. The large cell aggregates are disintegrated by the influence of shearing forces. But if we consider only the start of the agglutination reaction we can neglect the disintegration of flocs. In this case the integration of Eq. (8) in the range between n_o and n_t yields

$$\frac{1}{n_t} - \frac{1}{n_o} = k^+ \cdot t. \tag{9}$$

After multiplication of Eq. (9) with n_o we can write

$$\frac{n_o}{n_t} = 1 + k^t \cdot t \cdot n_o,$$

where n_o, the number of cells, is a constant value. If we combine it with k^+, we obtain

$$\frac{n_o}{n_t} = 1 + k \cdot t, \tag{10}$$

where n_o and n_t are substituted by Eq. (6). So the velocity of cell agglutination can be described simply by Eq. (10) and can be determined accurately by monitoring the optical density of the cell suspension

$$\left(\frac{E_0}{E_t}\right)^3 = 1 + k \cdot t = \tilde{n}. \tag{11}$$

In the plot of $(E_o/E_t)^3$ versus time a straight line arises. The slope of this line corresponds to k. This constant represents the velocity of the growth of cell agglutinates. For example $k = 1 \cdot \min^{-1}$ means that on average one cell is bound to another cell or to an agglutinate per minute.

The determination of only one rate constant k means a simplification of the agglutination kinetics. The whole agglutination process consists of a number of consecutive steps, and each step has its own rate constant. k is an average over the various steps of the agglutination process.

Though many considerations used for the derivation of Eq. (11) are not valid in our experiments, the transformation of the registrated extinction according to Eq. (11) yields a straight line over a wide range of time.

4.3 Determination of the Velocity of Cell Agglutination

Over the last 20 years many papers have been published which describe different optical methods suitable for quantitative determinations of cell agglutination (Bennert 1976; Beug and Gerisch 1972; Boehringer 1984; Hoffa 1980; Kaneko et al. 1975; Kenneth and Burger 1973; Krzywanek et al. 1977; Linnemans et al. 1976a, b; Maca and Hovack 1974; Ofek and Beachey 1978; Oppenheimer and Odenkrantz 1972; Rottman et al. 1974; Vlodavsky and Sachs 1975).

The described methods have various disadvantages, especially that of the poor reproducibility of the experimental data. So we can summarize that no method has been established which would allow the determination of the kinetics of cell agglutination under standardized conditions and with good reproducibility.

In our investigations we observed that the main problem consists in the agitation of the suspended cells in the cuvette. The stirrer must be adjusted exactly in the cuvette; it must not contact the walls or the bottom of the cuvette and the velocity of rotation has to be constant.

We solved this problem as illustrated in Fig. 4.1 (Flemming and Schulz 1987). We used a commercial glass cuvette measuring $10 \times 10 \times 40$ mm. The suspension was stirred by a disk with a diameter of 8 mm (1) that is fixed on a steel needle (2).

This needle is placed in two opposite corners of the cuvette at the bottom (3) and at the opening (4). The disk is adjusted in the middle of the cuvette. Thus the distance to all four walls is exactly 1 mm. No further equipment is necessary to stabilize this simple stirrer. A small gum (5) connected the steel needle to a synchronous motor (6) with 375 rotations per minute and the light beam (7) passed through the cuvette below the disk and beside the needle. The change in the optical density is registered continuously using a commercial spectral photometer.

The procedure is very simple. The cuvette is filled with 3 ml PBS (containing 10^{-4} M Ca^{2+}, Mg^{2+} and Mn^{2+}) and then a few microliters of concentrated cell suspension are added to the stirred buffer to obtain the desired optical density. Best results are obtained with an extinction of 0.75. After 1 min the agglutination is started by adding several microliters of lectin solution. As expected, parallel to proceeding cell agglutination, the optical density of the continously agitated suspension decreases steadily and reaches a constant value after 10 or 20 min. Usually, the cell agglutination does not start immediately, but after some seconds, so that an S-shaped curve is registered.

Figure 4.2 shows a typical agglutination curve. Erythrocytes were agglutinated by wheat germ agglutinin (WGA 8.3 µg/ml). The flocculation began very slowly after 20 s. As seen in Fig. 4.2 the decrease in the optical density remains constant during the following 2 min. This value ($\Delta E/\Delta t = 0.11$ min^{-1}), the tangent in the turning point, represents the so called apparent velocity of cell agglutination c (see below). During the fol-

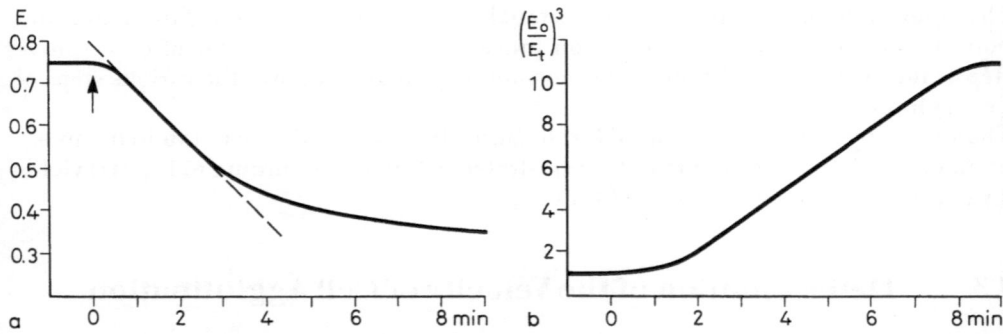

Fig. 4.2 Decrease in the optical density of suspended erythrocytes in correspondence to lectin-mediated cell agglutination
a) Decrease in the extinction (agglutination curve); b) transformation according to Eq. (11)

lowing minutes the decrease in E becomes slower and slower and the reaction ends after 8 min at $E = 0.34$.
In accordance with the theory the transformation of the extinction according to Eq. (11) yields a straight line in the periods between 1.5 and 7.0 min from the start of the reaction. The slope of 1.5 min^{-1} represents the "true" velocity of cell agglutination (binding of 1.5 erythrocytes to a single erythrocyte or to a floc per minute). At the end of the reaction the average number of erythrocytes per aggregate is $ñ = (0.75/0.34)^3 = 10.7$. Changes of the lectin concentration yield changes in v and in ñ.
The "apparent" velocity of cell agglutination can be determined very easily and very fast. These data are sufficient in many cases, e. g., determinations of lectin concentrations. But the "true" velocity of cell agglutination c has physical significane and it should be calculated in the case of basic investigations.

4.4 Determination of Lectin Concentrations

As expected, the velocity of cell agglutination depends on the concentration of the lectin in the stirred cell suspension. In accordance to Eq. (7) with increasing lectin concentration the velocity of cell agglutination also increases.
The maximum is reached when the number of lectin-occupied receptors is equal to the number of unoccupied ones ($\theta = 0.5$).
A further increase in the lectin concentration yields a reduction in the agglutination rate corresponding to a decrease in the term $\theta \cdot (1 - \theta)$ [see Eq. (7)] (Flemming et al. 1985 a).
Figure 4.3 illustrates the dependence of the cell agglutination on lectin concentrations. Two systems are described here: The agglutination of yeast cells (*Yarrowia lipolytica* EH 59/4) by *Ricinus communis* lectin (RCA$_{120}$) and the agglutination of erythrocytes (human A$_1$O) by wheat germ lectin (WGA). Unknown lectin concentrations are determined using Fig. 4.3 c, d as calibration curve.
As seen in Fig. 4.3 no agglutination occurs at a low concentration of lectins; increasing the concentration causes a rapid augmentation of the velocity of cell agglutination. In this range a very small change of lectin concentration induces a strong change in the

102

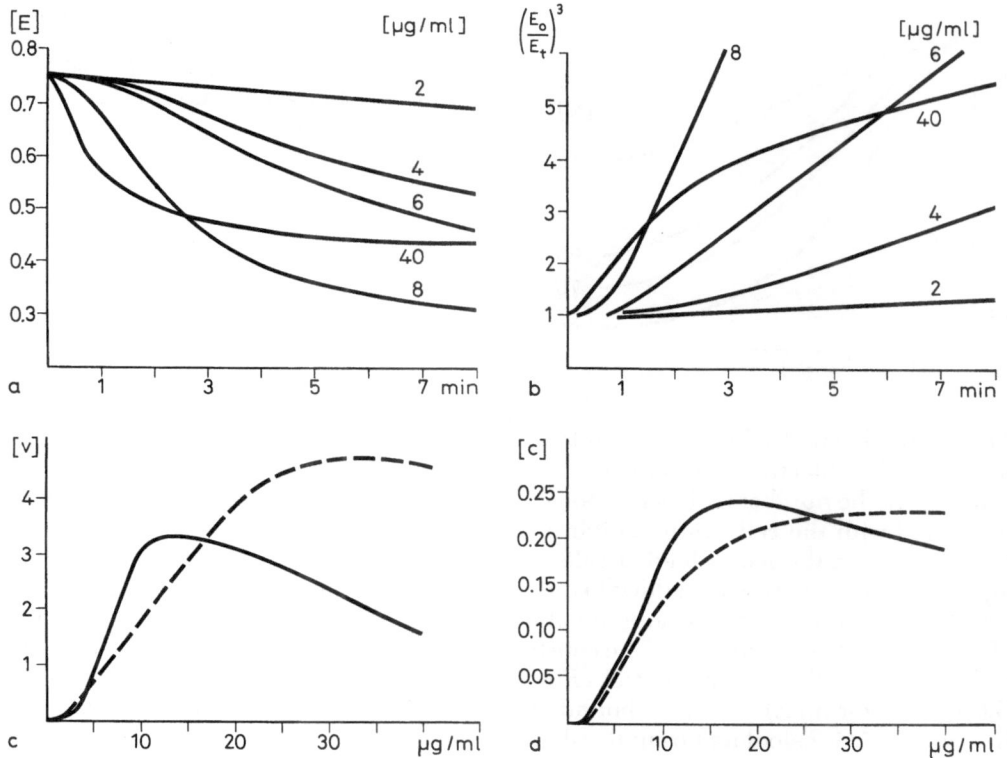

Fig. 4.3 Dependence of the velocity of lectin-mediated cell agglutination on the concentration.
a, b) Yeast cell agglutination (dotted line in c). d Erythrocyte agglutination. Yeast cells: $7.7 \times 10^6 \, \text{ml}^{-1}$; erythrocytes: $5.4 \times 10^6 \, \text{ml}^{-1}$. a) Agglutination curves registered by photometer; b) transformation of the curves according to eq. (11); c) dependence of the "true" agglutination velocity $v = (E_o/E_t)^3/t$ on the lectin concentration; d) dependence of the "apparent" agglutination velocity C (tangent in the turning point) on the lectin concentration

velocity of cell agglutination. Therefore, we are able to determine very exactly lectin concentrations in this region (error $<3\%$). The agglutination reaction is completed after 20 min (except at very low lectin concentrations). In the case of yeast cells, the average value for ñ (number of cells per agglutinate) reaches 11.8 and in the case of erythrocytes 17.5.

It is necessary to stabilize the stirrer. Small alterations in the intensity of agitation cause changes in v and in ñ. Using the cuvette, described previously, the registered values are of good reproducibility. The described method enables us to determine lectin concentrations with the same accuracy as enzymes through enzymatic reactions (for standardization of lectin preparations, see Bøg-Hansen et al. 1982).

The advantage of this simple and quick method is the exact determination of unknown lectin concentrations without any labeling of the lectin molecules. Lectin contents in raw materials and in commercial preparations are determined in less than 5 min. In Fig. 4.4 the different qualities of some commercial lentil lectin preparations are evident. Presumably, the low activity of some products is caused by extensive storage.

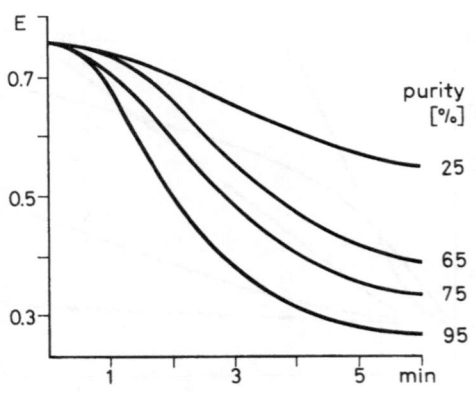

Fig. 4.4 Quality of some *Lens culinaris* lectin preparations. Agglutination of a distillery yeast *Saccharomyces cerevisiae* H 155 (1.1×10^7 cells/ml; 83 µg lectin/ml). The percentage was calculated using a calibration curve

This method can also be recommended for studying the temperature, pH, and long-term stability of lectins. We were also successful in estimating of the equilibrium constant K and the number n of lectin molecules bound to the cell. In this case the lectin is incubated with the cells until equilibrium is reached. Afterwards the supernatant is separated from the lectin-loaded cells. In a second step indicator cells with known agglutination properties are added to the supernatant containing the unbound lectin. Using a calibration curve, the amount of unbound lectin is calculated from the velocity of indicator cell agglutination. The constants are calculated by using Scatchard (1949) or Steck and Wallach (1965) plots. In Fig. 4.5 an example is given. Thus, 1.3×10^{-7} mol (7.8×10^{16} molecules) Con A are bound to 3.9×10^9 cells (2×10^7 molecules/cell). $K = 10^8 \text{mol}^{-1}$ is calculated from the slope of the curve.

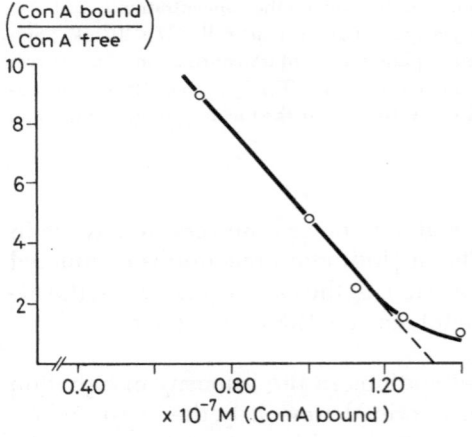

Fig. 4.5 Scatchard plot. Con A binding to yeast cells *Saccharomyces cerevisiae* H 155 (3.9×10^9 cells/l)

4.5   Microorganism Cells as "Lectin Test Substances"

For more than 100 years the ability of most lectins to agglutinate erythrocytes was described. It is also well known that erythrocytes of different origin or of different blood groups exhibit different agglutination properties (Roth 1978). Shore and Shore (1974),

Schnebli and Bächi (1975), and Singer and Morrison (1976) emphasized the influence of the metabolic state of erythrocytes on agglutination, i. e., the surface properties of the red blood cells of the same individuum may change from day to day.

It is also extremely difficult to stabilize erythrocytes over a long period in such a manner that the agglutination properties remain totally unchanged (see Turner and Liener 1975).

We investigated the lectin-mediated agglutination of microorganisms, which we now recommend as "test substances". Nonpathogenic microorganisms can be produced by mass cultivation in large amounts. Due to the strong stability of the cell wall (ca. 20 % of the dry weight) microorganisms can be lyophilized and stored without changing their agglutination behavior. In general they are also resistant to lytic substances.

Of course, no microorganism is expected to react with all lectins. But numerous nonpathogenic microorganisms with different carbohydrate molecules on the cell surface exist. Certainly, for each lectin with agglutinating ability many microorganism strains are available.

Strains of the yeast species *Yarrowia lipolytica* are of practical interest due to their ability to produce citric acid in considerable amounts. The cell wall of these yeasts contains galactose as well as mannose in terminal positions (Flemming 1988). We selected a highly sensitive reacting strain (EH 59/4), which can be agglutinated by several lectins including peanut and soybean lectin (both lectins agglutinate erythrocytes only after a special enzymatic pretreament). The agglutination curves are shown in Fig. 4.6. Unknown concentrations of the lectins can be determined using a calibration curve (see sect. 4.4).

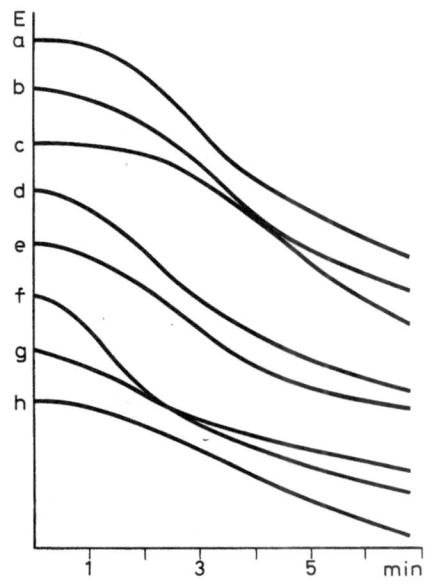

Fig. 4.6 Agglutination of yeast cells (*Yarrowia lipolytica* EH 59/4) by lectins with different specificities (7.7×10^6 cells/ml, lectin concentrations in µg/ml).
a) *Ricinus communis* (6.7); b) *Arachis hypogaea* (6.7); c) *Glycine max* (33); d) *Viscum album* (6.7); e) Con A (33); f) *Lens culinaris* (330); g) *Pisum sativum* (165); h) *Lathyrus odoratus* (165)

By testing the agglutination of cells on a slide, the lectin concentration may be reduced 10 or 20 % of that necessary for agglutination in the glass cuvette because no shearing forces destroy the flocs on the slide.

The cell wall of some microorganisms contains only one kind of carbohydrate. These

105

cells are suitable for selective determination and characterization of lectins. An excellent example is the yeast strain *Saccharomyces cerevisiae* H 155. The outer part of the cell wall consists of a highly branched mannan (the construction and composition of yeast cell walls have been reviewed by Ballou 1976, 1982; Cabib et al. 1982; Farkas 1979; Lampen 1968). The cells are very sensitive to all lectins with glucose-mannose specificity, but no reaction occurs with lectins of other specificities.

In the above described investigations the lyophilized (stabilized) microorganisms are comparable with a chromatogenic substrate in a photometric determination of an enzymatic reaction. The agglutination method allows an exact and reproducible determination of lectin concentrations.

4.6 Determination of Interactions Between Lectins and Saccharides

The lectin-induced agglutination of cells or precipitation of complex carbohydrates is usually inhibited by a simple monosaccharide, but for some lectins di-, tri-, or oligosaccharides are required.

The sugars are bound to the "active sites" of the molecules and prevent all further reactions of the lectin. Due to the simple chemical nature of the mono- or oligosaccharides, the interactions with lectins may be described by a simple equilibrium constant (association or inhibition constant, respectively). In some cases these interactions are determined immediately by measuring the change of the optical rotation dispersion during complex formation or by equilibrium dialysis. But in general for calculating the equilibrium constant the inhibition of lectin binding, agglutination, or precipitation has to be determined in a quantitative manner (the different methods are summarized by Liener 1986).

In the presence of a competitive carbohydrate the velocity of the lectin-mediated cell agglutination depends on the concentration and on the equilibrium constant of the lectin-sugar interaction.

The overall reaction can be written:

$$R_2L + 2C \rightleftarrows R + RL + 2C \rightleftarrows R + RLC + C \rightleftarrows 2R + LC + C \rightleftarrows 2R + LC_2 \quad (12)$$

$$2R + L + 2C$$

Where R denotes receptor molecules, L the lectin molecules with two binding sites, C the carbohydrate molecules and R_2L, RL, RLC, LC, LC_2 the corresponding complexes.

As seen in Eq. (12) the mixture contains a total of eight components.

The amount of each component is determined by the concentrations and the affinity of the components of the reaction. Unfortunately we can only measure the kinetics of the underlined step in Eq. (12) and not the equilibrium state. Using a calibration curve (dependence of the agglutination rate on the lectin concentration), the "free" lectin concentration may be determined. This should allow the estimation of the equilibrium constant. But most of the preconditions necessary for calculation (e. g., the number of receptors and their affinity constant must be known, only one type of receptors may exist) are not fulfilled. Therefore, an immediate determination of the equilibrium constant is difficult or impossible.

106

The inhibition of cell agglutination by simple saccharides depends on the ratio of the equilibrium constant and the sugar concentration. From Eqs. (13) and (14) the unknown inhibitory constant (K^-) of a carbohydrate (C^-) can be determined easily by comparison of the inhibition by a well characterized saccharide (C^+) with a known equilibrium constant (K^+).

$$\frac{[L] \cdot [C^+]}{[LC^+]} = K^+ \qquad \frac{[L] \cdot [C^-]}{[LC^-]} = K^- \qquad \frac{[L]}{[LC^+]} = \frac{K^+}{[C^+]} \qquad \frac{[L]}{[LC^-]} = \frac{K^-}{[C^-]} \quad (13)$$

Provided the inhibition by the well characterized carbohydrate (C^+) is equal to the inhibition of the unknown carbohydrate (C^-), we can write

$$\frac{[L]}{[LC^+]} = \frac{[L]}{[LC^-]} \quad \text{and} \quad \frac{K^+}{[C^+]} = \frac{K^-}{[C^-]} \quad \text{resp.} \quad K^- = K^+ \cdot \frac{[C^-]}{[C^+]} \quad (14)$$

Using Eq. (14) the unknown equilibrium constant is determined by comparison of a sugar with a known equilibrium constant.

Figure 4.7 shows the inhibition of *Ricinus communis* lectin-induced yeast cell agglutination by galactose and galactose-containing saccharides.

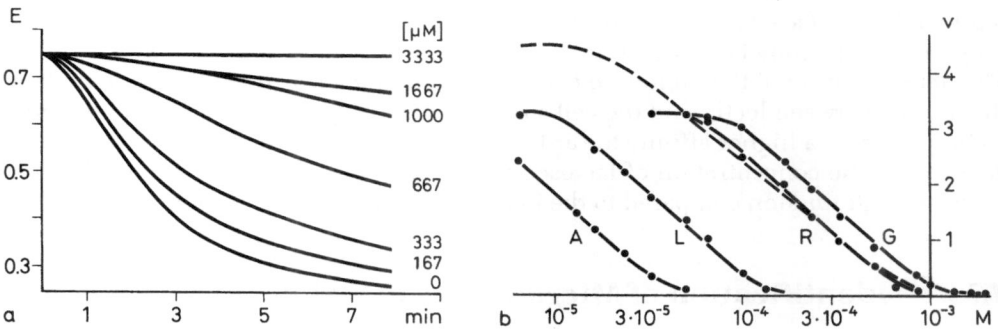

Fig. 4.7 Inhibition of lectin-mediated cell agglutination by competitive carbohydrates. Yeast cells (*Yarrowia lipolytica* EH 59/4; 7.7×10^6 cells/ml, *Ricinus* lectin 16.7 µg/ml).
G = Galactose, L = lactose, R = raffinose, A = gum arabic (Mw ~ 1300 per subunit which contains one galactose molecule in the terminal position). a) Agglutination curves at different concentrations of galactose. b) Velocity of cell agglutination in dependence on the concentrations of the competitive carbohydrates (logarithmic scale); the *dotted line* represents the inhibition of erythrocyte agglutination by lactose

Using the same cell type and the same lectin concentration, as expected, the shape of all "inhibitory curves" is exactly the same (see Fig. 4.7 b). They differ with regard to their position on the abscissa only. Liener (1986) noted that the inhibition of *Ricinus* lectin by lactose is 7.5 and by raffinose 1.5 times stronger than by galactose. As expected, the differences in the positions of the curves in Fig. 4.7 b are equal to the logarithm of 7.5 and 1.5 (0.88 and 0.17), i. e., in agreement with Eq. (14) these differences on the abscissa are equal to the differences of the logarithm of the equilibrium constants. Using a calibration curve (velocity of agglutination, depending on the concentration of a saccharide with a known inhibition constant), only one experiment should be necessary to estimate the unknown equilibrium constant.

As illustrated in Fig. 4.3 c the velocity of cell agglutination reaches a maximum at a lectin concentration of 14 µg/ml. We observed no alteration of the position of this maximum at different galactose concentrations, and the velocity of agglutination is generally decreased by all lectin concentrations. So it is impossible to induce cell agglutination with concentrations of galactose of 5 mM (even at an extremely high lectin concentration). This unexpected behavior is explained by Eq. (12): by increasing the carbohydrate concentration the equilibrium is shifted to the right side of the equation, and by increasing the lectin concentration more and more receptors are occupied and no free receptor molecules are available to form a bridge between two cells. In both cases the velocity of cell agglutination is diminished.

Normally, complex carbohydrates are precipitated by lectins. However, at low lectin or carbohydrate concentrations this reaction does not take place. But the agglutination method enables us to determine the interactions between lectins and high molecular carbohydrates. Figure 4.7 b illustrates the strong inhibition of *Ricinus* lectin-mediated agglutination by gum arabic, a branched polymer of galactose, rhamnose, arabinose, and glucuronic acid with $M_w \sim 250,000$.

As described in the previous section, the agglutination of microorganisms can be easily standardized. Therefore, these cells are very suitable for determining of the inhibitory constant of the lectin-sugar interaction. It is possible, however, to use other cells, e.g., red blood cells, (see Fig. 4.7 b). Their disadvantage is the alteration of agglutination properties over a long period of time.

The determination of the inhibition of cell agglutination also allows the estimation of the affinity between lectin and the cell surface receptor. As visible in Fig. 4.7 b *Ricinus* lectin possesses a higher affinity towards red blood cells than toward the investigated yeast cells. The concentration of lactose must be five times higher to reduce the erythrocyte agglutination compared to the inhibition of the yeast cell agglutination.

4.7   Identification of Microorganism Strains

Lectins have been shown to be a helpful tool in differentiation of bacteria (Köhler and Prokop 1967 a, b, c; Schneeweiss and Prokop 1967). The review of Pistole (1981) summarizes the first works in this field. Also, Allison (1986), Allen and Connely (1980), Cole et al. (1984), Craft et al. (1987), Curtis et al. (1987), Doyle and Keller (1984), Mayer et al. (1975), Niewerth et al. (1987), Uhlenbruck et al. (1985), Wagner (1982), and Weir (1980) used lectins successfully for grouping of bacteria strains. Schottelius (1982 a, b, c) demonstrated the possibility of characterizing *Leishmania* ssp. of different origin using lectins.

Methods generally used in the taxonomy of microorganisms permit the determination of genus and species. But it is difficult to distinguish between different strains of one species. All the "lectinological" methods described in the literature cited above allow a grouping of these microorganism strains into certain groups (e.g., sero groups) only. However, different strains of one group are not distinguished.

Usually, no further classifications are required for diagnosis. But in the field of biotechnology it is neccessary to characterize the selected microorganism strain accurately because only this strain exhibits the desired production properties. Other strains of the same species and also of the same sero-group can have drastically different properties. The problem of characterizing and identifying microbial production strains and of differentiating of those from other related strains of the same species is as yet unsolved.

Microbial production strains are generally characterized by their specific capabilities (e. g., generation of products, growth rate). The determination of these properties is time-consuming and the methods used are imperfect.

Microbial strains of one species, which cannot be distinguished by conventional methods, i. e., only with difficulties, may be differentiated quickly and accurately by determining their specific agglutination kinetics (Flemming et al. 1985b, 1988, 1989). Under standard conditions each strain shows specific kinetics of cell agglutination. The strains to be tested are cultivated under standard conditions. We recommend batch cultures in shaking flasks. The microorganisms should be harvested in the early stationary state. The cells are washed and suspended in buffer. The microorganims are agglutinated by adding a lectin solution; the velocity of cell agglutination is then determined by measuring the optical density as described above. The investigated strains are identified or distinguished by their specific agglutination curves. The method is not time and material consuming and can be used in routine examinations.

The enormous number of brewery, distillery, wine, and baker's yeast strains are classified in the species *Saccharomyces cerevisiae* (Barnett et al. 1983). The outer part of the cell wall of all these yeast strains is composed of a highly branched mannan (see sect. 4.5). Many of these strains are indistinguishable. Recently, Kersten et al. (1988) reported the impossibility to differentiate production strains of this species using common serological methods. Furthermore, Ballou (1982), Gabriel-Bruneau and Guinet (1984), Rademacher (1983), Stahl et al. (1983) and Zahn (1983) were unsuccessful in their investigations.

Unexpectedly, yeast strains of one species differ very much in their agglutination behavior[1]. Figure 4.8 shows the agglutination curves of the yeast strains which we have investigated (Flemming et al. 1989).

As expected, all investigated strains are agglutinated by Con A and the glucose-mannose-specific lectins from *Lens culinaris* (LCA), *Pisum sativum* (PSA) and *Lathyrus odoratus* (LOA) (Meinhold et al. 1979). In Fig. 4.8 a the strains are arranged according to their increasing agglutination rate. But Fig. 4.8 b, d, e illustrate that the established arrangement observed for Con A has no validity for other lectins. We did not observe a general rule concerning the agglutinability of the strains in dependence on the lectin. For understanding the lectin action it is notable that most, but not all yeast strains are agglutinated stronger at pH 4.5 than at pH 7.2. This effect was also observed in examinations of the Con A mediated agglutination of yeast cells (Flemming et al. 1985 a, Janson and Paktor 1977).

If the investigated strains exhibited nonidentical agglutination curves we considered them to be nonidentical. In the case of the same curves of two strains one cannot reliably conclude that they are "identical". Here, a second agglutination is necessary using another lectin or changed conditions (pH, concentration) for agglutination. At high agglutination velocities different strains show similar agglutination curves. Therefore

[1] The Research is supported by grants of the brewery Neubrandenburg (Nordbräu, Getränkekombinat Neubrandenburg, FRG)

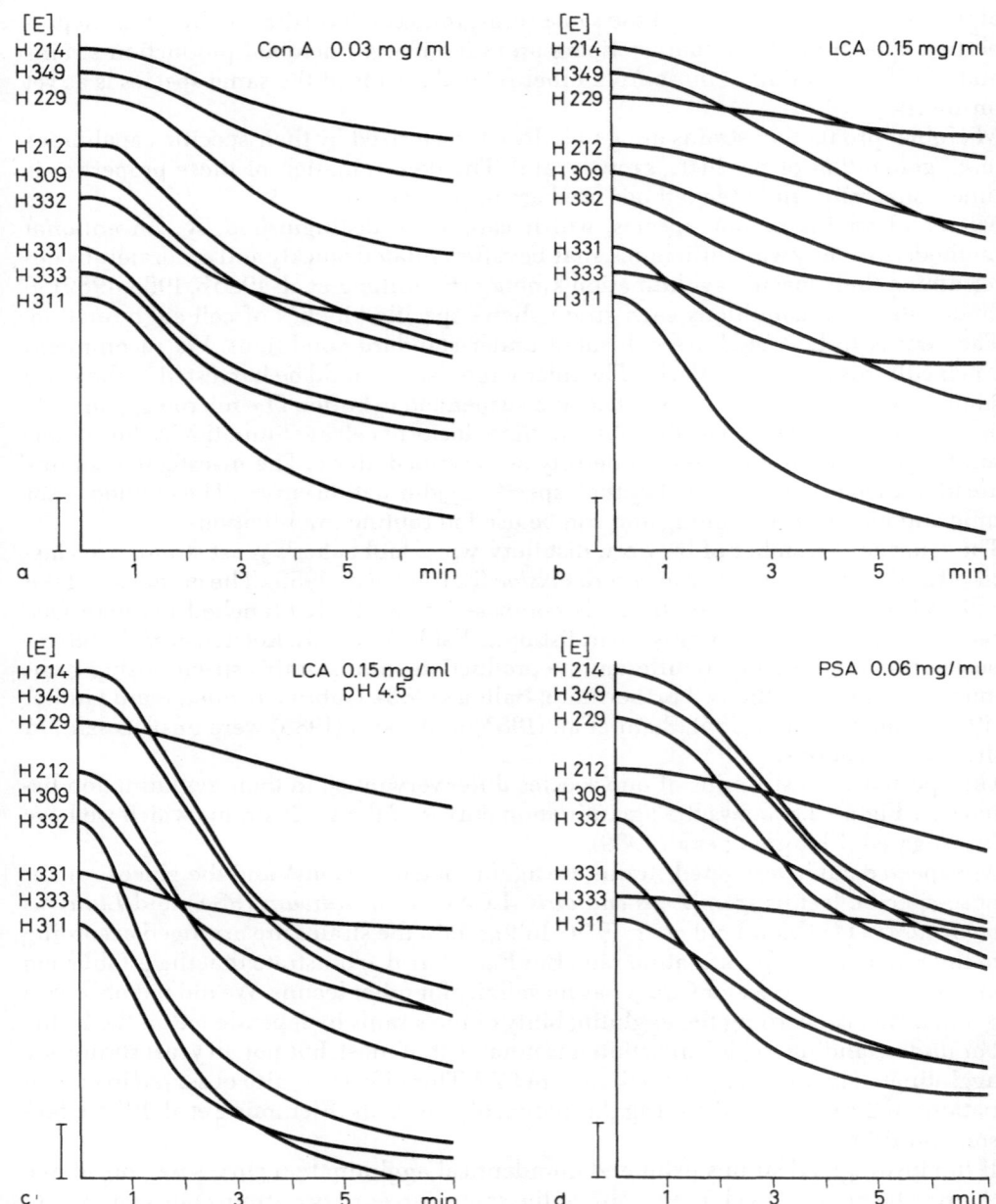

reaction conditions should be chosen which induce cell agglutination with low or medium velocity. Then the differences in the investigated strains can be seen best.

We also investigated the immune sera induced yeast cell agglutination (anti *Saccharomyces cerevisiae* antisera against the strain H 309). As illustrated in Fig. 4.8 f the immune sera possess only weak agglutination activities (an increase in the concentra-

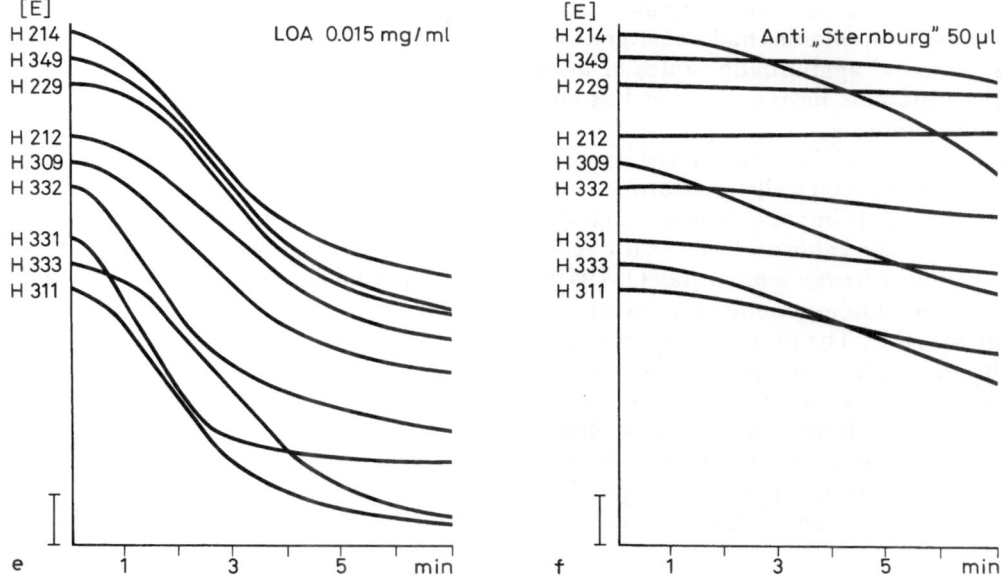

Fig. 4.8 a–f Agglutination curves of brewery yeast strains *(Saccharomyces cerevisiae)*. The culture collection number of the strains is given beside the curves

tion does not increase the agglutination rate). Various strains are not agglutinated by the sera (see Wegewitz 1986).

Some bacteria strains of the species *(Acetobacter methanolicus)* bind lectins but the cells failed to agglutinate. The bacteria are surrounded by a capsular polysaccharide. By heating the suspended cells the capsular shell is diminished. The so treated cells are agglutinated by lectins and the differentiation and characterization of the investigated strains is possible (Baeker 1985).

Summarizing many investigations over the last years we can say: due to their strong agglutination power, their identity and homogeneity in some molecular properties and their availability, lectins are a convenient tool to identify strains of microorganisms. Provided the lectins agglutinate cells, they are much more suitable than immune sera.

Using the agglutination procedure, the properties of a cell population are described by a curve containing three types of information: the time difference between the addition of the lectin and the beginning of the agglutination, the velocity of agglutination, and the volume of the generated flocs. For quantification of lectin binding (using FITC- or POD-labeled lectins) the cell properties are described by a single measuring point only. Therefore, we were not able to differentiate related strains using fluorescence-labeled lectins, but these strains exhibited different agglutination properties.

4.8   Characterization of Mutants

We investigated more than 300 nonpathogenic strains of microorganisms, but we observed no correlations between classification and agglutination rate. The agglutination rate seems to be strain-specific and not genus- or species-specific.

Studying mutants of microorganism strains we observed altered agglutination properties when the colonies had a changed morphology. Unexpectedly, we also found drastically altered agglutination rates of mutants with an obviously unchanged cell wall. These mutants have a more or less changed biochemical pathway (Flemming and Sattler 1985).

Normally, fungi do not grow as a homogeneous cell suspension (concerning the structure of fungal cell walls, see Burnett 1978; for the reaction of fungi with lectins, see Pistole 1981). It is impossible to characterize fungi strains using the agglutination procedure described above. However, the spores exhibit very good agglutination properties. The fungus *Trichoderma reeseii* QM 9414 is a potent producer of cellulase. UV- and nitrosomethylurea-generated mutants and fusants are characterized by the agglutination of spores. The fungi were grown on a special agar which induces sporulation. After filtration and washing the spores are agglutinated by addition of Con A or *Lens culinaris* lectin (LCA) (Flemming et al. 1988 b). The spores are agglutinated at low pH only (acetate buffer, pH 4.5). The strains are arranged in Fig. 4.9 a according to the increasing agglutination velocity. By using *Lens culinaris* lectin (Fig. 4.9 b) instead of Con A, the arrangement is changed. Obviously, the velocity of cell agglutination is a function of the specific combinations of lectins and strains or mutants resp.

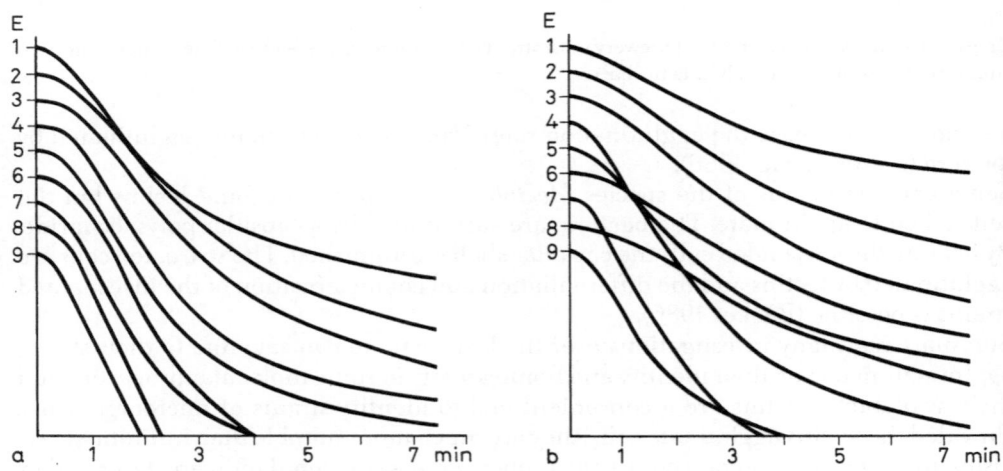

Fig. 4.9 Agglutination curves of mutants and fusants of fungi spores (*Trichoderma reeseii* QM 9414) a) Con A 33 µg/ml; b) LCA 67 µg/ml. No 6 = Strain, Nos. 1, 2, 3, 4, 8 mutants; Nos. 5, 7, 9 fusants

As mentioned in the previous section, the agglutination using only one lectin does not allow a reliable differentiation between the mutants in each case. In general, the lentil lectin differentiates better than Con A in characterization of mutants and in strains.

We already reported on the identification of mutants of yeast strains (Flemming and Sattler 1985) and bacterial strains (Baeker 1985). The agglutination method enables us to select mutants and prove their stability. It is not certain that all mutants can be detected, but we are sure to find all forms which exhibit changed properties concerning the composition of the cell wall or the biochemical pathway.

We investigated the purity and homogeneity of microorganism production strains in type culture collections. Single cells are isolated and cultivated separately. Unexpectedly we could distinguish two different "agglutination types" in apparently homogeneous strains. We believe that by the frequent transfers of strains from agar tube to agar tube, mutants have evolved and partially enriched and displaced the original strain. Therefore, we recommend the storage of selected microorganism strains in liquid nitrogen or lyophilic cultures.

4.9 Biological Control of Fermentation Processes

The stability of a microorganism population or a cell line is an essential precondition for all fermentation processes. Infections, mutations of the production strain, or technical disturbances diminish the generation or the quality of products.

Bacteria of the species *Acetobacter methanolicus* are well known to produce gluconic acid. In some fermentation experiments we observed the substitution of the selected strain by another strain of the same species. Using taxonomic methods both strains are indistinguishable. Under special cultivation conditions, the second strain produces only small amounts of the desired gluconic acid. The cells of this strain are agglutinated by Con A, whereas the production strain fails to agglutinate at 165 µg Con A/ml. After taking a sample from the fermenter, this agglutination method enables us to check the purity of the production strain immediately.

Figure 4.10 illustrates the infection of the culture and the complete displacement of the production strain by the Con A-agglutinating strain in the fermenter during the following days. The amounts of both strains are calculated from a calibration curve.

Kersten et al. (1988) investigated the reactions of immune sera with distillery and brewery yeast strains. The authors were able to distinguish "culture" and "wild type" yeasts, but it was impossible to differentiate distillery and brewery yeasts. When a brewery and a distillery plant are in the same vicinity, an infection of the brewery strain with the distillery strain and vice versa is possible.[1] Both yeast strains are agglutinated by glucose-mannose-specific lectins. If the strains are grown on glucose as carbon source, they exhibit similar agglutination rates but if cultivated on beer only the cells of the brewery yeast are strongly agglutinated at low concentrations of *Lens culinaris* lectin. This fact enables us to determine the amount of each component in a mixed population (see Fig. 4.10c; Flemming et al. 1988a).

To study microorganism populations in long-term fermentations it is necessary to isolate single cells. These were grown to colonies on agar plates and further cultivated in shaking flasks. Finally, the agglutination properties of the different samples were investigated.

Recently, we studied a strain of the hydrocarbon utilizing yeast *Candida maltosa* in a long-term fermentation process. We isolated four "agglutination types" all belonging to the species *Candida maltosa*. All forms are indistinguishable with common methods. The amount of the major component in the mixture was 75%. Surprisingly, no isolate showed the same agglutination behavior as the original strain. This example demonstrates that lectins may be an excellent tool to study mutations and evolutionary processes.

[1] These investigations are supported by grants of the brewery Neubrandenburg (Nordbräu, Getränkekombinat Neubrandenburg, FRG)

113

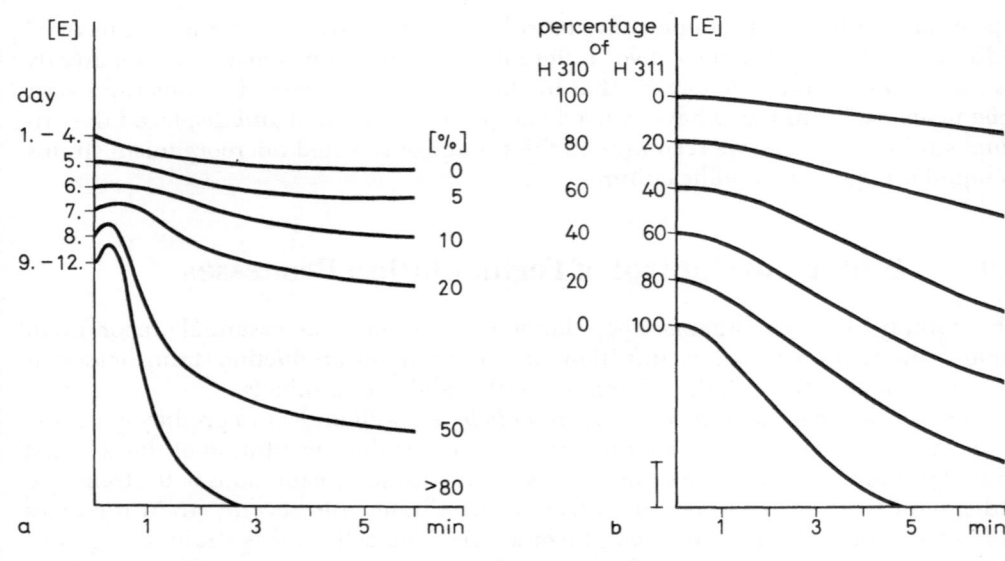

Fig. 4.10 Agglutination curves of a mixture of two microorganism strains.

a) Displacement of the production strain *Acetobacter methanolicus* by a second strain belonging to the same species (optical density of the suspension E = 1.5 instead of E = 0.75, 165 µg Con A/ml). b) Agglutination curves of a mixture of a distillery and a brewery yeast (33 µg LCA/ml). c) Calibration curve for determination of the percentage of each yeast strain in the mixture

4.10   Agglutination and Metabolic State

It is well known that growth conditions influence the properties of the cell membrane and cell wall (Roth 1980). In the study of a population of *Acanthamoeba castellanii* Linnémanns et al. (1976b) and Spies et al. (1975) noted a cell agglutination by Con A in the logarithmic multiplication phase only. Dazzo et al. (1979) and Hrabak et al. (1981) described a dependence of the trifoliin-induced agglutination of *Rhizobium trifolii* on the growth phase of the bacteria. However, qualitative alterations in the composition of the cell wall are exceptions. In general, if the growth conditions of the cell population are varied, the composition of the cell surface is altered in a quantitative manner only.

As expected, the rate of the lectin induced cell agglutination depends on the physiological state of the cell population (Flemming et al. 1986a). In the case of strain identification this dependence has been shown to be a disadvantage. Therefore the microorganism strains or cell lines have to be cultivated under identical conditions. On the other hand, the influence of the metabolic state on the agglutination enables us to study and describe the "physiological state" of a cell population. Consequently, a possibility is given for monitoring fermentation processes. These investigations are successful only at low or medium velocity of cell agglutination. Alterations of the properties of cells are not detectable at agglutinations with a high velocity.

Figure 4.11 illustrates the change of the *Lens culinaris* lectin-induced cell agglutination during the cultivation of yeast cells in a shaking flask.

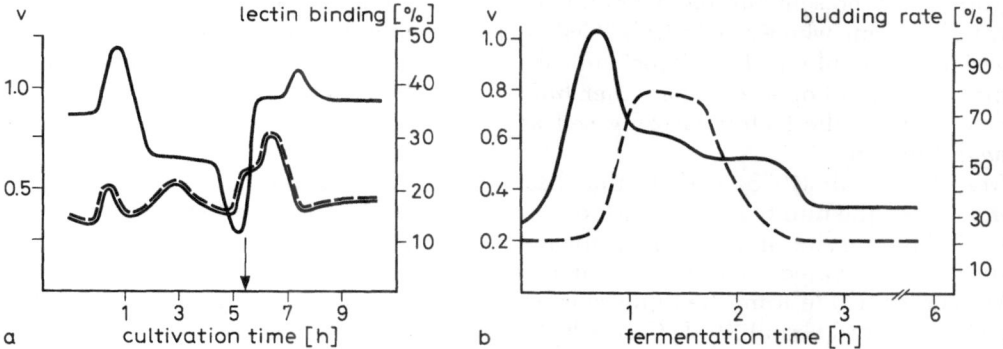

Fig. 4.11 Dependence of the agglutination rate of yeast cells *Saccharomyces cerevisiae* H 155 on the "physiological state" of the culture.
a) Shaking flasks culture (1.1×10^7 cells/ml, 83 µg FITC-LCA/ml), velocity of the cell agglutination,... binding of the lectin. b) Synchronous culture (1.1×10^7 cells/ml, 50 µg LCA/ml) velocity of the cell agglutination; ... budding rate

The arrow in Fig. 4.11 a indicates the end of the logarithmic growth phase. A very strong increase in the velocity of cell agglutination is observed in the period between the logarithmic and stationary phase. Within 40 min the agglutination rate increases more than fourfold. In this experiment we used fluorescein-labeled *Lens culinaris* lectin and studied both the lectin binding and the cell agglutination. Obviously, no close correlation exists between binding and agglutination. But in general, the agglutination rate increases if more lectin is bound to the cells.

In Fig. 4.11 b the agglutination behavior of a synchronous culture in a fermenter is demonstrated (Flemming et al. 1986 a). At the starting point a small amount of glucose is added, enough for a single cell replication only. The velocity of cell agglutination is increased already after 20 min, but an increase in the budding rate is observed after 40 min. From these results we can conclude that the alteration in the agglutination behavior reflects changes in the physiological state of the cells.

In the study of the fermentation of bakers yeast, Repp et al. (1989) observed strong correlations between the activity of alcohol dehydrogenase and the rate of cell agglutination.

At present only a few details are known about the processes wich make the cell more or less agglutinable. However, the possibility of measurement will lead to a better understanding of the phenomenon.

Our further investigations concern the monitoring of fermentation processes in biotechnology. The quantitative determinations of the interactions of lectins with microorganisms will provide more information on the "physiological state" of the population as well as on disturbances of the process.

4.11 Mechanism of Lectin-Induced Agglutination of Microorganisms

The mechanism of cell agglutination has been investigated extensively over the last years, but at present our understanding is far from being complete. Presently we agree with Nicolson, who wrote in 1974 "Cell agglutination is a complex process involving a variety of the physical and biochemical parameters or factors. These various factors may or may not oppose one another, but agglutination should occur in any given cell system when the factors favoring cell agglutination outweigh the factors opposing agglutination."

Moreover Roth (1978) added "that besides the biochemical nature of the lectin molecules, the number of cell surface bound lectin molecules and the mobility of the lectin binding sites in the plain of the membrane, other factors are of importance for the lectin mediated cell agglutination as well."

We can conclude from the numerous investigations (Granth and Peters 1984; Kahn 1982; Ketis and Granth 1983; Ochoa 1979; Rutishauser and Sachs 1974, 1975a, b; Singer et al. 1972, 1973) that the phenomenon of differential agglutination by lectins of microorganism strains as well as the differential agglutination of normal and transformed cells is not completely understood, and it is not possible to predict whether or not a given lectin agglutinates a certain cell line.

The mechanism of mammalian cell agglutination is totally different from the mechanism of microorganism cell agglutination. Mammalian cells are enveloped by a membrane and the lectin receptors are complex glycoconjugates. These receptors possess mobility and may change their position in the membrane. Nicolson (1974) and Vlodavsky and Sachs (1975) proposed that the mobility of lectin receptors is an essential precondition for cell agglutination. In their experiments a dramatic inhibition of lectin-mediated cell agglutination was reported after chemical fixation, low temperature, and other factors which reduce the receptor mobility (but the lectin binding was not reduced).

In contrast to mammalian cells microorganisms posses a strong cell wall with rigid receptors (e. g., Janson and Paktor 1977). These receptors cannot be redistributed (and cannot form a cluster or cap). It seems that a certain flexibility of these receptors is necessary however, for cell agglutination.

Kahn (1982) investigated the agglutination of some bacterial strains by lectins and observed that lectin binding was either nonspecific or positively cooperative. Agglutination was observed only in combinations which showed positive cooperative binding. Lectins binding to the same carbohydrates did not necessarilly bind to the same microorganisms.

In the study of the agglutination of bacteria by lectins we were successful in the differentiation of three steps in the reaction: lectin binding, rearrangement of receptors, and cell agglutination. The strain *Acetobacter methanolicus* MB 135 is agglutinated by glucose-mannose-specific lectins. After addition of the lectin solution to the suspended cells the registered optical density increases during the first part of the reac-

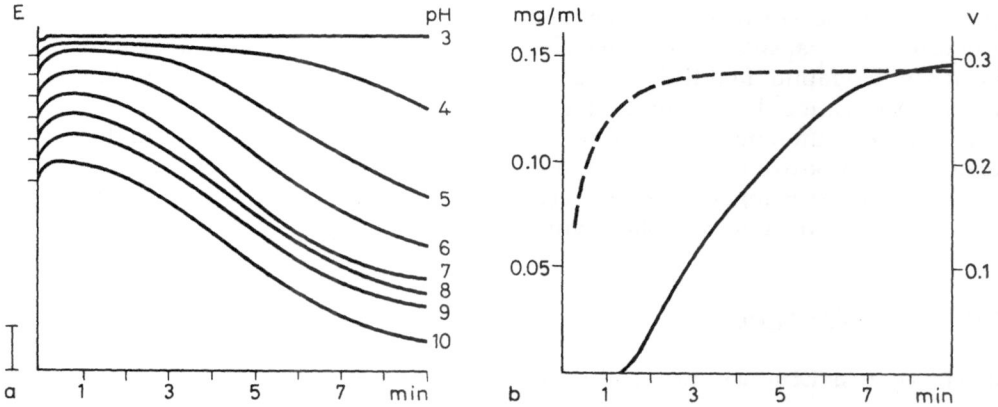

Fig. 4.12 Agglutination of bacteria cells *(Acetobacter methanolicus)* by *Pisum sativum* lectin.
a) agglutination curves at different pH values; b) velocities of lectin binding and cell agglutination velocity of
agglutination, ... amount of lectin bound to the cells, pH 4.5

tion. The expected cell agglutination (decrease in the extinction) starts after some minutes. As illustrated in Fig. 4.12 a the retardation of agglutination increases with reduction of the pH value.

Baeker (1985) demonstrated that the investigated *A. methanolicus* strains are surrounded by a polysaccharide capsule. This capsule is not seen under a light microscope and absorbed no light in the agglutination test. But after addition of a lectin the capsular polysaccharide is cross-linked. This precipitation like reaction leads to turbidity of the capsule and to the observed increase in the extinction. (The isolated capsular polysaccharide, a mannan, is also precipitated by lectins).

Due to the retardation of agglutination at low pH values we were able to determine separately the velocities of lectin binding and cell agglutination. As seen in Fig. 4. 12 b, lectin binding is a fast reaction but a time difference exists between lectin binding and the start of the agglutination reaction. The optical density of the suspended cells immediately increases when 165 µg *Pisum sativum* lectin is added to 1 ml cell suspension ($\sim 10^{10}$ cells/ml). After equilibrium is reached 145 µg are bound to the cells. Already 90 % of this amount is bound during the first 1.5 min. After this time the agglutination starts with a very low velocity and reaches a constant value very late, after 7 min. This time is necessary for a rearrangement of the receptor molecules of the lectin loaded receptors. Only lectin molecules which are bound to the newly ordered polysaccharide chains are able to form a bridge between two cells. At low pH the velocity of the rearrangement of the lectin receptors is decreased but not the velocity of lectin binding.

The "rearrangement hypothesis" is supported by further observations. Some *A. methanolicus* strains bind FITC-labeled lectins but fail to agglutinate. These cells exhibit an increase in the optical density after the lectin is added, but show no agglutination. The strains bind lectin molecules but a rearrangement of the capsular polysaccharide chains seems to be impossible.

Strains of *Pasteurella haemolytica* are agglutinated by wheat germ lectin. We investigated native and formalin-treated cells and observed no agglutination during a period of more than 20 min, but after this time the cells started to agglutinate immediately and with a high velocity. Presently we are unable to predict whether or not bacteria are

117

agglutinated. Recently, Grimmecke (1989 pers. commun.), determined the complete structure of the capsular polysaccharide of *A. methanolicus* MB 58 and a r-mutant of the strain. He found no differences in the composition. Both strain and mutant bind *Ricinus* lectin, but only the r-mutant is agglutinated.

Generally the differences between the calculated agglutination curves and the registered ones may be explained by the inhibition of the start of the reaction due to the time necessary for a rearrangement of the receptors. As seen in Figs. 4.6 and 4.8 this retardation depends on the nature of the lectin and on the cell type.

4.12 References

Allen P, Conelly M (1980) Interaction of lectins with *Neisseria gonorrhoeae*. Can J Microbiol 26:468–471

Allison R (1986) Lectins in diagnostic: a review. Med Lab Sci 43:369–376

Baeker R (1985) Untersuchungen zur Reaktion von Lektinen und Immunglobulinen mit methanolas-similierenden Bakterienstämmen. Diplomarbeit, Karl-Marx-Univ, Leipzig

Ballou C (1976) Structure and biosynthesis of the mannan component of the yeast cell envelope. Arch Microb Physiol 14:93–158

Ballou C (1982) Yeast cell wall and cell surface. In: Strathern J, Jones E, Broach J (eds) The molecular biology of the yeast *Saccharomyces*, metabolism and gene expression. Cold Spring Harbor Monogr Ser, New York, pp 335–360

Barnett J, Payne R, Yarrow D (1983) Yeasts: characteristics and identification. Cambridge Univ Press, London New York

Bennert C (1976) Untersuchungen zur Auslösung und Hemmung der Thrombozytenaggregation mit dem photometrischen Blättchenaggregationstest (PAT II). Inaugural-Dissertation, Univ Frankfurt/Main

Beug H, Gerisch G (1972) A micromethod for routine measurement of cell agglutination and dissociation. J Immunol Methods 2:49–57

Boehringer, Firmenschrift (1984) Thrombozytenaggregationstest. Boehringer biochemica-dienst Nr. 60

Bøg-Hansen T, Breborowicz J, Franz H (1982) Report on the proposals for an international working party on standardization of lectins for diagnosis. In: Bøg-Hansen (Ed) Lectins: biology, biochemistry, clinical biochemistry, vol 2. De Gruyter, Berlin New York

Burnett J (1978) Fungal walls and hyphal growth. In: Burnett J, Trinci A (eds) Symp Mycol Soc April 1978, Lond

Cabib E, Roberts R, Bowers B (1982) Synthesis of the yeast cell wall and its regulation. Ann Rev Biochem 51:763–793

Cole H, Ezzell J, Keller K, Doyle R (1984) Differentiation of *Bacillus anthracis* and other *Bacillus* species by lectins. J Clin Microbiol 19:48–53

Craft D, Chengappa M, Carter G (1987) Differentiation of *Pasteurella haemolytica* biotypes A and T with lectins. Vet Rec 12:393

Curtiss R, Pearce C, Pollack J, Murchison H (1987) Isolation and characterization of mutants of *Streptococcus* mutants using selective removal of wild type cells. Acta Microbiol Pol 36:3–15

Dazzo F, Urbano M, Brill W (1979) Transient appearance of lectin receptors on *Rhizobium trifolii*. Curr Microbiol 2:15–18

Doyle R, Keller K (1984) Lectins in diagnostic microbiology. Eur J Clin Microbiol 3:4–9

Farkas V (1979) Biosynthesis of cell wall of fungi. Microbiol Rev 43:117–144

Flemming C (1988) Determination of the velocity of lectin mediated cell agglutination – a new method for identification of microorganism strains of one species. 10th Int Lectin Conf 3.–8. July 1988, Prague, Czechoslovakia

Flemming C (1989) Lektine in der Biotechnologie. Wiss Z Karl-Marx-Univ Leipzig 38:342–348

Flemming C, Sattler K (1985) Identifizierung von genetisch abweichenden Formen eines Produktionsstammes durch Bestimmung der Agglutinationsgeschwindigkeit. 2nd Symp DDR–UdSSR 30 Sept–5 Oct 1985, Reinhardsbrunn

Flemming C, Schulz HJ (1987) Vorrichtung zur Bestimmung der Agglutinationsgeschwindigkeit von suspendierten Zellen. DDR WP C 12 N 300 805 8

Flemming C, Gabert A, Flemming I (1985 a) Wechselwirkung zwischen Lektinen und Mikroorganismen. 1. Bestimmung der Agglutinationsgeschwindigkeit durch Extinktionsmessungen: Agglutination von Hefezellen (*Saccharomyces cerevisiae* H 155) durch Concanavalin A. J Basic Microbiol 25:493–501

Flemming C, Flemming I, Schädlich H, Seliger B, Gabert A, Ringpfeil M, Sattler K (1985b) Verfahren zur Identifizierung und Charakterisierung von Mikroorganismen. DDR WP C 12 Q/274 207·8

Flemming C, Seliger B, Flemming I, Hilger U, Ringpfeil M, Heinritz B, Rogge G (1986a) Verfahren zur Kontrolle und Steuerung mikrobieller Fermentationsprozesse. DDR WP C 12 N 296 640 4

Flemming C, Repp H, Seliger B, Tenckhoff V (1986b) Verfahren zur Kontrolle und Steuerung von mikrobiellen Stoffwandlungsprozessen. DDR WP C 12 N 296 641 2

Flemming C, Flemming I, Babel W, Pöhland D, Miethe D, Schröter J, Schulz H, Laube K, Wesenberg J (1988a) Verfahren zur Bestimmung der quantitativen Zusammensetzung von mikrobiellen Mischpopulationen. DDR WP C 12 Q 312 944 3

Flemming C, Kerns G, Dalchow E, Flemming I, Lohse U (1988b) Verfahren zur Identifizierung und Charakterisierung von Pilzstämmen. DDR WP C 12 Q 323 950 4

Flemming I, Flemming C, Laube K, Wesenberg J, Zahn G (1989) Characterization and differentiation of brewing yeasts. Interlec 11, Tartu Estonia, June 4–9. 1989, (Abstr)

Gabriel-Bruneau SM, Guinet RM (1984) Antigenic relationships among some *Candida* species studied by crossed-linked immunoelectrophoresis: taxonomic significance. Int J Syst Bacteriol 34:227–236

Granth C, Peters M (1984) Lectin-membrane-interactions: information from model systems. Biochim Biophys Acta 779:403–422

Hoffa J (1980) Magnetic stirrer for sample container of photometric analyzer. US Patent 4:227 815

Hrabak E, Urbano M, Dazzo F (1981) Growth-phase-dependent immunodeterminants of *Rhizobium trifolii* lipopolysaccharide which bind Trifoliin A, a white clover lectin. J Bacteriol 148:697–711

Janson V, Paktor J (1977) The effect of temperature on Con A mediated agglutination of cells with rigid receptors. Biochim biophys Acta 467:321–326

Kahn L (1982) Nonspecific and cooperative binding of lectins to microorganisms. Physiol Chem Phys 14:3–7

Kaneko J, Hazatsi H, Ukita T (1975) A quantitative assay for Concanavalin A– and *Ricinus communis* agglutinin-mediated agglutinations of rat ascites hepatoma cells. Biochim Biophys Acta 392:131–140

Kenneth N, Burger M (1973) The relationship of Concanavalin A binding to lectin-initiated cell agglutination. J Cell Biol 59:134–142

Kersten R, Zahn G, Wesenberg J, Schade W (1988) Serologische Differenzierung von Hefen. – Herstellung und Austestung von Antiseren gegen Kulturhefen und Wildstämme. Lebensmittelindustrie 35:121–124

Ketis N, Granth C (1983) Time dependent lectin binding to isolated receptors in model membranes. Biochim Biophys Acta 730:359–368

Köhler W, Prokop O (1967 a) Agglutination von Streptokokken der Gruppe C durch ein Agglutinin aus *Helix pomatia*. Z Immunforsch 133:50–53

Köhler W, Prokop O (1967 b) Agglutinationsversuche an Streptokokken mit dem Phytagglutinin aus *Dolichos biflorus*. Z Immunforsch 133:171–175

Köhler W, Prokop O (1967 c) Untersuchungen über die Anwendbarkeit von Anti B_{gal} und Anti H_{per} zur Streptokokkenidentifizierung. Z Immunforsch 133:491

119

Krzywanek H, Grun H, Breddin K (1977) Thrombocytenfunktionstests. In: Engelhardt A, Lommal H (eds) Diagnostik hämorrhagischer Diathesen. Verlag Chemie, Weinheim

Lampen J (1968) External enzymes of yeast: their nature and formation. *Antonie v. Leeuwenhoek* J Microbiol Serol 34:1–18

Liener IE (1986) The lectins: properties, functions and applications in biology and medicine. In: Liener IE, Sharon N, Goldstein IJ (eds) Academic Press, Lond New York Orlando

Linnemans W, Wiersema P, Spies F, Elbers P (1976a) A kinetic model for cell agglutination. Exp Cell Res 101:184–190

Linnemans W, Spies F, De Ruyter De Wildt Th, Elbers P (1976b) Kinetics of cell agglutination. A quantitative assay of Con A mediated agglutination of *Acanthamoeba castellani*. Exp Cell Res 101:191–201

Maca R, Hovack J (1974) Improved method for quantitation of Concanavalin A induced agglutination. J Natl Cancer Inst 52:365–367

Mayer H, Schlecht S, Gromska W (1975) Reactivity of LPS from various *Salmonella* SR and R mutants with Concanavalin A. Zentralbl Bakteriol Hyg I Abt Orig A 233:327

Meinhold I, Gabert A, Wünsche L (1979) Agglutinationsverhalten von Hefen mit Lektinen. Z Allg Mikrobiol 19:741–744

Nicolson G (1974) The interactions of lectins with animal cell surfaces. Int Rev Cytol 39:89–190

Niewerth B, Lämmler C, Blobel H (1987) Reactions of lectins with animal pathogenic *Streptococci* of the serological group C and S. Zentralbl Veterinärmed 34:206–210

Ochoa J (1979) The mechanisms of lectin-mediated cell agglutination. Pathol Biol 27:103–113

Ofek J, Beachey EH (1978) Mannosebinding and epithelial cell adherence of *Escherichia coli*. Infect Immun 22:247–254

Oppenheimer S, Odenkrantz J (1972) A quantitative assay for measuring cell agglutination. Agglutination of sea urchin embryo and mouse teratroma cells by Concanavalin A. Exp Cell Res 73:475–480

Pistole T (1981) Interactions of bacteria and fungi with lectins and lectinlike substances. Ann Rev Microbiol 35:85–112

Rademacher K. (1983) Strukturuntersuchungen an Mannanen der Hefe. Biochem Physiol Pflanzen 178:307–312

Repp H, Tenckhoff V, Bayer C, Gründig B, Flemming C (1989) Biosignale in Gärprozessen. Acta Biotechnol (in press)

Roth J (1978) The lectins, molecular probes in cell biology and membrane research. Fischer, Jena

Roth J (1980) The use of lectins as probes for carbohydrates-cytochemical techniques and their application in studies on cell surface dynamics. Acta Biochem (Suppl) 22: 113–121

Rottman W, Walther B, Hellerquist C, Umbreit J, Roseman A (1974) A quantitative assay for Concanavalin A mediated cell agglutination. J Biol Chem 249:373–380

Rutishauser U, Sachs L (1974) Receptor mobility and the mechanism of cell-cell binding induced by Concanavalin A. Proc Natl Acad Sci USA 71:2456–2460

Rutishauser U, Sachs L (1975a) Receptor mobility and the binding of cells to lectin coated fibers. J Cell Biol 66:76–85

Rutishauser U, Sachs L (1975b) Cell to cell binding induced by different lectins. J Cell Biol 65:247–257

Scatchard G (1949) The attractions of proteins for small molecules and ions. Ann NY Acad Sci 51:660–672

Schnebli H, Bächi T (1975) Reaction of lectins with human erythrocytes. I. Factors governing the agglutination reaction. Exp Cell Res 91:175–189

Schneeweiss B, Prokop O (1967) Zur Agglutinabilität einiger Salmonellen gegenüber zwei Protektinen. Acta Biol Med Germ 19:615

Schottelius J (1982a) The identification by lectins of two strains of *Trypanosoma cruzi*. Z Parasitenkd 68:147–154

Schottelius J (1982b) Lectin binding strain specific carbohydrates on the cell surface of *Leishmania* strains from the Old World. Z Parasitenkd 66:237–247

Schottelius J, Goncalves da Costa S (1982) Studies on the relationship between lectin binding carbohydrates and different strains of *Leishmania* from the New World. Mem Inst Oswaldo Cruz Rio de Janeiro 77:19–27

Shore B, Shore V (1974) The interaction of Concanavalin A with sheep erythrocytes. Biochim Biophys Acta 373:313–326

Singer J, Morrison M (1976) Effect of metabolic state on agglutination of human erythrocytes by Concanavalin A. Biochim Biophys Acta 426:123–131

Singer S, Nicolson G (1972) The fluid mosaic model of the structure of cell membranes. Science 175:720–731

Singer J, Vekemans F, Lichtenbelt J, Hesselink F, Wiersema P, (1973) Kinetics of flocculation of latex particles by human gamma globulin. J Colloid Interface Sci 45:608–614

Spies F, Linnemans W, De Ruyter De Wildt TH, Hax W (1975) Growth phase dependent Concanvalin A agglutinability of *Acanthamoeba*. Cytobiologie 11:65–86

Stahl M, Kües M, Esser K (1983) Flocculation in yeast, an assay on the inhibition of cell aggregation. Eur J Appl Microbiol Biotechnol 17:199–202

Steck T, Wallach D (1965) The binding of kidney-bean phytohemagglutinin by Ehrlich ascites carcinoma. Biochim Biophys Acta 97:510–522

Turner R, Liener J (1975) The use of glutaraldehyde treated erythrocytes for assaying the agglutinating activity of lectins. Anal Biochem 68:651–653

Uhlenbruck G, Lütticken R, Bäz K, Böhmer G (1985) Serologische Kreuzreaktionen von Gruppe B Streptokokken. Immun Infekt 13:276–282

Vlodavsky J, Sachs L (1975) Lectin receptors on the cell surface membrane and the kinetics of lectin-induced cell agglutination. Exp Cell Res 93:111–119

Wagner M (1982) Agglutination of bacteria by a sialic acid-specific lectin of the snail *Cepaea hortensis*. Acta Histochem 71:35–39

Wegewitz B (1986) Untersuchungen zur Identifizierung von Hefestämmen durch spezifische Reaktionen von natürlichen Antikörpern mit der Zelloberfläche. Dissertation, Akad Wiss DDR, Leipzig

Weir D (1980) Surface carbohydrates and lectins in cellular recognition. Immunol Today 1:45–51

Zahn G, Kersten R, Schade W, Wesenberg J (1983) Die Bedeutung serologischer Verfahren für die Differenzierung von Hefen in der Gärungs- und Getränkeindustrie. Lebensmittelindustrie 30:411–414

Index

Advances in Lectin Research, Vol. 1

1. Auflage 1988, 188 S., 44 Abb., 3 Tab., 2 Schemata,
L6 = 17 × 24 cm, Leinen, 95,00 DM
Bestell-Nr. 534 475 5/Franz, Lectinologie 1, engl.

ISBN 3-333-00204-3 Verlag Gesundheit
ISBN 3-540-17972-0 und 0-387-17972-0 Springer-Verlag

The new series "Advances in Lectin Research" covers the rapid development in the research on lectins, major tools in many biological, biochemical, immunological and clinical laboratories.
Advances in Lectin Research's aim is to publish comprehensive revues written by leading experts in the field. Supported by an international editorial board, the series editor will see to it that these reviews will be of interest not only to the "true lectinologists". But also to any scientist and technican in biology and medican working with lectins in his research.

The first volume contains the following rewievs:
The Ricin Story · Hartmut Franz
Preparation of Plant Lectins · Harold Rüdiger
Structure and Function of Leguminosae Lectins.
Edilbert van Driessche
Illustrations of Lectin-producing Plants (I) · Christa Beurton,
Renate Israel, Hartmut Franz

Interested areas: lectinologists, immunologists, biologists, clinical chemists, biotechnologists

Advances in Lectin Research, Vol. 2

1. Auflage 1989, 112 S., 35 Abb., 28 Tab., L6 = 17 × 24 cm,
Leinen, 98,00 DM
Bestell-Nr. 534 632 0/Franz, Lectinology 2, engl.

ISBN 3-333-00316-3 Verlag Gesundheit
ISBN 3-540-18961-0 und 0-387-18961-0 Springer-Verlag

Volume 2 contains the following rewievs:
Lectins as Mitogens · C. A. K. Borrebaeck, R. Carlsson
Viscaceae Lectins · H. Franz
Mechanism of Action of Ricin and Related Toxic Lectins on the Inactivation of Eukaryotic Ribosomes · Y. Endo
Effects on Gut Structure, Function and Metabolism of Dietary Lectins. The Nutritional Toxicity of the Kidney Bean Lectin · A. Pusztai
Potential Participation of Tumor Lectins in Cancer Diagnosis, Therapy and Biology · H.-J. Gabius

Interested areas: lectinologists, immunologists, biologists, clinical chemists, biotechnologists

Advances in Lectin Research, Vol. 3

1. Auflage 1990, 172 S., 22 Abb., 9 Tab., 16 Diagramme,
L6 = 17 × 24 cm, Leinen, 98,00 DM
Bestell-Nr. 534 801 9/Franz, Lectinology, 3, engl.

ISBN 3-333-00441-0 Verlag Gesundheit
ISBN 3-540-51240-3 und 0-387-51240-3 Springer-Verlag

Volume 3 contains the following rewievs:
Biochemical Properties of Vertebrate 14K Beta-Galactoside-Binding Lectins · K. Kasai
Sialic Acid-specific Lectins · M. Wagner
Egg Lectins of Invertebrates and Lower Vertebrates: Properties and Biological Function · A. Krajhanzl
Illustrations of Lectin-producing Plants (II) · Christa Beurton, Renate Israel, H. Franz

Interested areas: lectinologists, immunologists, biologists, clinical chemists, biotechnologists